Charles Dixon

Annals of Bird Life

a year-book of British ornithology

Charles Dixon

Annals of Bird Life
a year-book of British ornithology

ISBN/EAN: 9783337095420

Printed in Europe, USA, Canada, Australia, Japan

Cover: Foto ©berggeist007 / pixelio.de

More available books at **www.hansebooks.com**

ANNALS OF BIRD LIFE

A Year-Book of British Ornithology

BY

CHARLES DIXON,

AUTHOR OF "RURAL BIRD LIFE," "EVOLUTION WITHOUT NATURAL SELECTION,"
"OUR RARER BIRDS," "STRAY FEATHERS FROM MANY BIRDS," ETC. ETC.
PART AUTHOR OF "A HISTORY OF BRITISH BIRDS."

"Many study Nature in the house, and when they go out of doors cannot find her."
AGASSIZ.

WITH ILLUSTRATIONS.

LONDON: CHAPMAN AND HALL,
LIMITED.
1890.
[*All rights reserved.*]

CHARLES DICKENS AND EVANS,
CRYSTAL PALACE PRESS.

PREFACE.

ONE of the greatest charms attached to the study of Ornithology is to keep a careful register day by day and month by month of the various movements and habits of birds. The recording of these facts is absolutely the best apprenticeship an aspiring student can serve; it teaches him to see things for himself, not to rely on statements made by others; it gets him into the best way of investigating Nature's secrets; and above all, it furnishes him with an endless store of fascinating information relating to the economy of wild creatures. Diaries nowadays are often filled with too much trivial, not to say worthless matter—with stupid thoughts, and chronicles of trashy incidents, which would be put to much better use were they more often made the receptacle for the story told by Nature and her works. The country is full of charm—its interest never wanes; and that district becomes the most interesting which is the most carefully and intelligently explored.

In the following pages I have endeavoured to place before the reader a few of the stirring incidents and vivid scenes of bird life during the cycle of the year. It has been my object to convince him that interesting information respecting the feathered tribes may be gleaned through every month of the twelve. It has been my constant endeavour to impress on the student the fact that the birds he observes and studies are animated by Life precisely the same as his own, and to allow these feathered creatures that important share of intelligence which their high mental qualities demand. Too long have birds been regarded as mere machines; yet their little lives, far from being the prosaic, automatic existence we have been so accustomed to suppose them, are full of poesy and intellectual fire. To appreciate thoroughly the habits and the ways of birds we must never lose sight of their mental attributes.

I have so arranged the present volume that the student may readily acquaint himself with what is going on among the birds during every month of the year. I make no pretensions to completeness; the subject is too vast and varied to be exhausted in a single volume. In many cases the dates given can of necessity only be approximate ones—an allowance of a few days either way must often be made for differences of latitude, state of the weather, and other local influences.

Should this little volume serve to increase the interest taken in Natural History, or prove an incentive to the keeping of local records of the ways and movements of birds, the labour of compiling it from the note-books filled during twenty years of field and forest errantry will not have been altogether a vain or a fruitless one.

CHARLES DIXON.

6, Ingatestone Terrace,
    Warren Road, Torquay.

CONTENTS.

Part I.—Spring.

CHAPTER I.
                                                        PAGE
THE GLORIES OF THE SPRING 1

CHAPTER II.
AMONG THE BIRDS IN SPRING 10

CHAPTER III.
STRANGERS OF THE SPRING 38

CHAPTER IV.
BIRD ORNAMENTS AND TOURNAMENTS 52

CHAPTER V.
SPRING-TIME ON THE MOUNTAINS 61

CHAPTER VI.
OUR FEATHERED ENGINEERS 70

CALENDAR FOR SPRING 87

Part II.—Summer.

CHAPTER I.
THE WONDERS OF THE SUMMER 93

CHAPTER II.
AMONG THE BIRDS IN SUMMER 102

CHAPTER III.
FEATHERED FRAUDS 124

CHAPTER IV.
THE WAYS OF BIRDS 134

CHAPTER V.
AMONG THE WHEAT 143

CALENDAR FOR SUMMER 154

Part III.—Autumn.

CHAPTER I.
THE BEAUTIES OF THE AUTUMN 160

CHAPTER II.
AMONG THE BIRDS IN AUTUMN 169

CHAPTER III.
STRANGERS OF THE AUTUMN 193

CHAPTER IV.
MIXED CONGREGATIONS 231

CHAPTER V.
WHERE THE MIGRANTS GO 240

CHAPTER VI.
PARAGRAPHS ON PLUMAGE 249

CALENDAR FOR AUTUMN 256

Part IV.—Winter.

CHAPTER I.
THE TERRORS OF THE WINTER 263

CHAPTER II.
AMONG THE BIRDS IN WINTER 272

CHAPTER III.
SOME VISITING CARDS 296

CHAPTER IV.
SOME BIRDS OF THE WINTER 309

CHAPTER V.
BEDTIME 322

CALENDAR FOR WINTER 335

ANNALS OF BIRD LIFE

Part I.—Spring.

CHAPTER I.

THE GLORIES OF THE SPRING.

Come, gentle Spring! ethereal Mildness, come;
And from the bosom of yon dropping cloud,
While music wakes around, veil'd in a show'r
Of shadowing roses, on our plains descend.

SPRING! What a rich feast for the naturalist is embodied in that one little word! It means to him the awakening, as it were, into life of many forms of animals and birds and plants. It means unwonted activity in every domain of Nature, a season of glad sounds and pleasant sights. All creatures love the spring-time; they vie with each other in making merry on Nature's birthday. The groves re-echo songs of gladness from countless feathered throats; insect life in endless variety sallies forth with incessant hum; flowers spring up like magic from every wood and hedgerow; each tree and bush shows signs of returning vitality; and animals that have been lying snugly dormant through the wintry blasts come out into the world again and flit and leap about for joy. In early spring the whole realm of Nature seems like a restless sleeper about to awaken after the long winter's night. Every day the buds on trees and hedges expand more and more, and many a tiny shoot may be detected amongst the carpet of dead leaves. The hazel bushes are gay with tasselled ornaments; primroses cast their pale faces upwards from the mossy banks, snowdrops and daffodils carpet the woodland glades. The woods, which in winter looked gray and net-like, appear much more dense when viewed from a distance, and brown tints steal imperceptibly over them as the millions of tiny buds swell out.

Wherever we may wander during the months of spring, we never need be at a loss for instruction and amusement. In the cool, green woods, when May has carpeted them with bluebells; by the stream, as the softest shades of greenery steal over the alder trees; on the moors, whose wide expanse is the chosen haunt of many interesting birds; along the lanes; in the fields; by the sea-shore—everywhere the glories of spring are unfolding. Here we may muse on the secrets of the life around us; here we may study its philosophy, and enjoy its manifold pleasures.

One of the greatest triumphs which natural science has achieved in modern times is the recognition it has succeeded in obtaining for the mental capabilities of the lower animals. The old school of philosophers, whose views on Mind were extremely limited and bigoted, has given place to a newer and more enlightened class of observers. Birds, for instance, are no longer regarded as mere automatic machines, governed by mysterious impulses all vaguely classed under the convenient term of "instinct," but as creatures endowed with mind, with mental powers very similar to those which control the movements of man himself. At no other period of the year, perhaps, are these various mental powers so well displayed as in the spring-time. For instance, what enormous powers of memory birds call into action in performing their long journey from Africa, and other remote regions, to their summer quarters in this country! What passion and jealousy animate them in the pairing season; what a large amount of imitation, reason, and forethought are required in the all-important task of selecting a site for the nest, and then in building the structure itself! How much more interesting and fascinating, too, does the study of birds become when, instead of classing all this brain-work as mysterious "instinct," we watch the progress of the little Mind which prompts these actions, and note the endless variation of the method by which these mental powers are used! Take the subject of Migration first. The tiny leaves are just bursting from the buds on the birch trees in this grand old Yorkshire coppice; the ferns are beginning to uncurl their fronds deep down in the damp mossy nooks between the big boulders of millstone grit. Spring is creeping rapidly over the valleys and clothing every twig and spray with delicate green. Suddenly, as if by magic, the Chiffchaff appears. Scores and hundreds of them may be heard *chiff-chaffing* from the birch and alder trees, and even from the long bilberry wires and heath that in some places half conceals the rocks. No man saw these birds arrive; silently they make their *début* in their summer quarters, journeying to them in the night when all is still and the road is safe. Whence have these little feathered wanderers come? They are all the way from Northern Africa; from the oases in the Great Desert; from the groves of Morocco and Fez and the country of the lawless Touareg. They have crossed the Straits of Gibraltar, passed along the coasts of Portugal, Spain, and France, and over the stormy English Channel; then two hundred miles further still, nearly the length of England, to the old familiar coppices in this Yorkshire valley. Think of the magnitude of such a journey; fifteen hundred miles of flight for a pair of little wings almost as delicate as gossamer, supporting a body which would go inside a big thimble! Think of the little mind encased in this feathered casket; the recalling to memory of old familiar landmarks on the way; the eye for detail; the knowledge of locality brought into action between the date-palms of Algerian oases and the bilberry wires and birch trees of Yorkshire! And remember that this identical Chiffchaff has come right back again to the spot where it built its nest last year, and is about to make a fresh one not half-a-dozen yards away from the very bush which shielded its previous home! How very much more fascinating is the philosophy of a rational migration, than to assume that the Chiffchaff is prompted to fly hither it feels not how, it knows not why! The Chiffchaff has had to learn its way like any other traveller; and has been taught many of the tricks of the journey by former experience. Then remember the Chiffchaff is only a unit in the great migrating army of birds; and that there are others which come from more distant regions still—the Swallows, for example, which traverse sultry Africa from end to end and perform a journey of six thousand miles twice every year of their lives.

But we have only touched upon the very margin of this fascinating subject yet. Watch our little Chiffchaff hopping about the alder bushes by the stream. Note the intelligence conveyed in its little eye, now expressive of alarm, now of boundless trust, questioning you, reading your very thoughts. It is perfectly amazing what an amount of expression can be seen in the eyes of a bird—an unerring index to the mental powers at work in the brain. Then comes the courtship of our stranger Chiffchaff. How boldly the cock-bird chases his mate amongst the tree-tops; or how softly he whispers to her down among the bilberry wires; and how coyly she receives his advances! No two birds pay court the same; each love-making has its individual peculiarities, prompted by the little reasoning mind behind the feathers. Jealousy is rife even in such a tiny creature, and he will fight furiously with any rival male that intrudes upon his lady-love or seeks to alienate her affections. Then, the wedding over, we may watch the Chiffchaffs searching out a site for their nest. In and out of the wires and heather stems, up and down the alder bushes, amongst the bluebells and the polypoddies, every nook is surveyed, and its advantages discussed, until the mutual choice is made at last. Next comes the building of the nest; memory, imitation, and reason each playing an important part in the course of its construction. The merry little male goes out to seek for grass and moss and feathers, pausing every now and then to sing his simple song, and bringing them bit by bit to his mate, who does the greater part of the building. Gradually the home is made, a beautiful example of bird architecture, half-domed, and snugly hidden away behind a cluster of fern and heath and grass below a birch tree on the hillside. Day after day goes by; the cock-bird sings more joyously than ever; there is true love in this old Yorkshire coppice, and where there is that there is happiness. Their home has been finished this three days; and now the pearly eggs are laid morning after morning until six of the precious speckled treasures take up the greater part of the nest. The pair of Chiffchaffs are brimful of joy; you can see it in their expressive eyes, and observe it in their every action. Then comes the long period of incubation; male and female sharing the loving task between them; until finally their patience and their self-sacrifice are rewarded by the appearance of six wee, tender, weakly nestlings. All through the lengthening spring days the faithful little parents feed and tend their brood, watch over them and care for them until they are able to quit the nest and begin life for themselves.

There is something in the very air of spring conducive to musing on the secrets of wild life. For me it is a season of wonderment and awe, and always inspires thoughts that bear upon the higher philosophy of the science I love. To watch the stirring pageant of life being marshalled into review order by that one grand solar force is pleasure indeed. Then right through the glorious spring-time the mental faculties of birds may be studied most advantageously, and note-books may be filled with charming facts bearing upon this fascinating subject. This is the season for the awakening of life, and the commencement of Nature's revels. The woods and fields are full of objects of interest; every day increases their abundance. No pen can do justice to the glories of the spring. In every direction life in countless varied forms is appearing. Let the naturalist revel in the feast which Nature so unstintingly provides. All the wild creatures of the woods and fields are his very own—his to study and observe, his to record their ways and movements, his to love, admire, and protect.

The pleasures of the naturalist who delights to chronicle the ever-changing phenomena of spring, and loves the birds and flowers and insects for their own sakes, and for the charming memories they recall of seasons past and gone, may be simple ones, yet, with all their simplicity, they never fail to yield true happiness to a contemplative mind, and furnish more real gratification than all the more costly enjoyments combined that wealth can bestow. Bounteous Nature showers her gifts with unsparing hand around us, yet her varied pleasures never satiate or tire the mind; and what is more, no calamity can rob us of the rich store of information we can glean in her pleasant pathways—once gathered, it remains an endless source of delight and recreative change. I firmly believe that if the charm of outdoor observation of Nature were more widely known, attended as it is with all the benefits of early hours, fresh air, and tranquillity of mind, the mentally distracted would seek relief by becoming naturalists; and the heated ball-rooms and wild dissipations of town life would lose their seductive attractions in the higher and the nobler and the healthier pursuit. I would not thrust upon such novices the dry details and the drier jargon of science, for such are apt to repel rather than convert; but I would teach them to watch the development of wild life in woods and fields, and to chronicle the ways of all wild creatures. Let us therefore do all we can to popularise the study of Natural History; to guide the younger generation to the wilderness rather than to the dancing rooms and theatres, for in doing so we elevate the mind to a far loftier standard, and save the body from many of the deadly perils with which nineteenth-century civilisation surrounds it.

Musing thus, we are apt to forget the ebbing away of time until reminded by the settling gloom and the noisy Blackbirds that night has come again. We must therefore reserve for our next chapter some of the most prominent and stirring incidents of bird life in the spring.

CHAPTER II.

AMONG THE BIRDS IN SPRING.

SPRING-TIME among the birds is replete with interest for the naturalist; and every day, nay, almost every hour, the feathered tribes are becoming imbued with ever-increasing animation and activity. Perhaps the first birds to feel and foretell the advent of spring are the Thrushes. At morn and even the speckled Song Thrush pipes his oft-repeated strains from the evergreens or the lofty and still leafless trees; whilst the Missel-thrush is occasionally heard, especially during the boisterous days of March. The sweet-voiced Blackbird is in full song during the very earliest days of spring, and his mellow music breathes the prophecy of coming life over bare hedgerows and leafless woods. The Starlings feel the influence of the changing year, and may be observed, with drooping wings and puffed-out plumage, sitting on the chimney-stacks or waterspouts warbling their curious song. Spring's magic influence spreads far and wide over all living creatures. Birds hitherto silent suddenly

<small>Missel-thrush ceases to sing, 16th April.</small>

begin to sing; flocks of birds separate; a Robin's nest in all its rustic beauty calls for our admiration; the Rooks begin to build, after much noisy deliberation, in the tops of the tall elm trees; and we often wonder how they can carry such operations on at all, so ceaselessly do the long branches sway and bend to and fro in the high March winds. In the first few days of spring those birds that build early are busy; and love, courtship, and marriage are their ruling impulses. Among many birds "house-hunting" is the order of the day. We see the comical little Blue-tits exploring every knot-hole and cranny, and deliberating and discussing the conveniences of each before finally choosing one for their nest. Others, more fortunate, return to the old familiar nest-hole, and jealously guard it from the intrusion of wandering strangers. Robins, Wrens, and Thrushes are all on the look-out for suitable nesting sites, and when once the great question is settled, they never wander far from the spot until the all-important duties of the year are over. As time goes on, and boisterous March has dried up the rills and fallows, and April comes once more with her fickle smiles of sunshine and her copious tears of rain, the migratory birds begin to make their appearance. First and foremost of these little strangers is the Wheatear, *Wheatear arrives, 25th* a bird that frequents the stony grounds, old *March.* quarries, downs, and sheep-walks. Almost simultaneously with his arrival we hear the monotonous notes of the Chiffchaff from the woods. Few birds call so incessantly as this little brown-coated Warbler; his voice is heard in never-changing tones until he leaves us again in autumn, or, to be more accurate, until the moult commences. It is a very singular circumstance that the Chiffchaff is without a song, especially when we know his congener, the Willow Wren, warbles most melodiously, and one which seems to show the Chiffchaff's much closer affinity with the larger and more brightly coloured Wood Wren, whose musical powers are also smaller than the Willow Wren's. In its habit of frequenting trees, and in the colour of its eggs—a most important character among this family of birds—the Chiffchaff is certainly more closely allied to the Wood Wren than it is to the Willow Wren. A few days later the latter little bird steals silently back from his winter retreat in Africa to his summer quarters in the woods, and gardens, and fields of our own country. Soon after his arrival his remarkably sweet and plaintive little song may be heard in almost every tree and bush, and a few weeks afterwards the pretty, semi-domed nest is built amongst the brambles and tall herbage on some mossy bank half-hidden by anemones and nodding bluebells, in which the female lays half-a-dozen tiny white eggs, speckled with pale reddish-brown.

Chiffchaff sings, 1st April.

Willow Wren arrives, 5th April.

One of the most marked characteristics of bird life in early spring is that it becomes more dispersed. In winter birds congregate in localities best able to supply them with food and shelter, but in spring they spread themselves over the entire country. Delicate little Wagtails run daintily over the clods of earth and along the furrows in the wake of the plough, eagerly catching the early insects or feeding on the larvæ which the share is exposing; the Skylarks and Yellow Buntings are back again on the grass fields; the Meadow Pipits leave the lowlands and seek the breezy moors where they delight to rear their young. Lapwings, Golden Plovers, and Dunlins desert the coasts and mud-flats and retire to inland mountain districts; the Ring-doves begin to engage in courtship, and shortly afterwards commence building; the homely Robins and noisy Blackbirds leave the vicinity of houses where they have lived through the winter and go back to their old haunts in the woods and hedgerows; the flocks of Linnets and Twites and Redpoles disband and seek their summer quarters on moors and commons and in the fir plantations on the hillsides. The same changes may be observed amongst waterfowl. Ducks and Coots and Moorhens return to the quiet ponds; the various species of Gulls desert localities where they lived in thousands all the winter and congregate at their usual breeding-places; whilst the great bird nurseries round the rocky coasts, which have been practically deserted hitherto, are once more full of their feathered tenants. From all parts of the surrounding seas, these sea-birds arrive in count-

Pied Wagtails migrating, 23rd March.

Ring-doves coo, 1st March.

Linnets disband, 24th March.

less hosts—Guillemots, Puffins, Razorbills, Kittiwakes, Gannets, and Fulmars, all moved by a common impulse to meet at the old familiar trysting-places. The Divers leave the sea and retire to inland lochs, as also do great numbers of Oystercatchers and Ringed Plovers.

<small>Puffins return to St. Kilda, 1st May.</small>

Right through the spring the migration of birds is going on apace. The army of Geese and Ducks and the hosts of Arctic Waders which were driven south to our coasts last autumn hurry home again to lands bathed in bright sunshine and gay with brilliant flowers, where summer comes quickly and without almost any warning sign, transforming the Arctic regions from howling wastes into verdant plains and valleys which now become a vast aviary of birds. We now miss the flocks of Redwings on the pastures, and the chattering Fieldfares are heard no more amongst the hawthorns; they have sped north to the birch and fir forests on the slopes of the Swedish fells. The pretty Snow Buntings have set out on their Arctic journey, some of them wandering probably even to the Pole. But if we lose many of our little feathered friends, which have been our daily companions since the autumn, there are others constantly arriving to take their place. From the end of March to the middle of May an almost incessant stream of birds is pouring into this country from sultry Africa, too hot and parched for them to live in now. By the middle of April we welcome back the Swallows and the Martins,

<small>Redwings all departed, 3rd April.</small>

<small>Sand Martin arrives, 15th April.</small>

<small>Swallow arrives, 18th April.</small>

and by the end of the month the Cuckoo's note rings clearly out from the woods and groves. This blithe harbinger of summer follows the spring-time northwards from his winter retreats in South Africa as that season gradually spreads over Central and Northern Europe. With the earliest dawn of spring in South Europe he makes his appearance, towards the end of March; but ventures no farther on his travels until the April sunshine woos the dormant vegetation back to life. Thus in England he does not arrive until towards the end of April; in Scotland a week later still; and in the Arctic regions not before the beginning of June. As with most of our summer migrants, the male Cuckoos are the first to appear, the females a few days later. Nor do these male birds utter their blithesome notes until the lady Cuckoos join them; consequently the Cuckoo's cry by no means proclaims the exact date of the bird's arrival. Many other migratory birds are similar in this respect. From the last week of April to the beginning of June, the Cuckoo is perhaps the most prominent bird of the woods and fields. His rich, loud, far-sounding notes echo from the wide expanse of woodland, from the trees in the meadows and the hedgerows, from the alders by the river-side, from the coppices and spinneys, from the moors, the mountains, and the heaths. With the voice of the Cuckoo some of the pleasantest moments of the whole year are inseparably associated. His happy notes are the

House-Martin arrives, 18th April.

wedding-bells for innumerable little songsters, now married and settled and enjoying their brief honeymoon amongst the cool green leaves and the early blossoms. His cry unerringly foretells the pageant of summer and heralds the triumphant appearance of Life in its endless forms. Every now and then the big, hawk-like Cuckoos chase and toy with each other as they fly from tree to tree, scolding and chattering as they go; and sometimes we are quite startled by their loud, hollow cry in the branches overhead, when, perchance, we are standing hidden amongst the nut bushes, knee-deep in bluebells and fern. The Cuckoo is bereft of all nest-building impulses, and, as every naturalist knows, selects the home of some other bird, and commits its egg and future offspring to the care of foster-parents. A gay Lothario is the Cuckoo, saddled with no parental cares, spending the entire summer in gadding about, enjoying life to the full, and wandering off back again to Africa before the first russet-and-yellow tints steal over the woods. Now the Nightingale, the Blackcap, the Garden Warbler, the Whitethroat, and other soft-billed birds also arrive in full force and take up their residence in the woods and gardens and hedgerows. By the end of April the Corncrake is back again amongst the meadow grass and growing corn, and his monotonous cry is heard all day long, and often best part of the night as he wanders from field to field in quest of his mate. Among the very latest birds to arrive

Warblers arrive, 15th April.

are the Goatsucker, the Turtle Dove, the Spotted Flycatcher, and the Swift. From all parts of the cool, shady undergrowth of briar and thorn in the woods and spinneys, the rich, wild song of the Nightingale sounds almost incessantly. What delightful melody indeed escapes from this little, brown-coated chorister—music which sounds like bubbling water, so pure and clear that it makes us wish we could by some refrigerating process freeze the song as it issues forth, and preserve it in all its loveliness until a season when gray skies and bare boughs take the place of warm May sunshine and vernal foliage. But perhaps, after all, the song is best in keeping with summer's birthday, when all Nature is in the first freshness of its youth. The Nightingale is no obtrusive musician—he loves the shadiest spots, where the sunlight only penetrates here and there in embroidered masses on the foliage, and where, when he flies, his red tail flicks conspicuously in the subdued light. High up among the golden-green shimmering leaves of the poplars the Wood Wren sings continuously; whilst among the thickets below Blackcaps, Whitethroats, and Garden Warblers hold their matchless concert. Round and round above the trees by the water-side the House Martins and the Swallows glide and circle and turn in quest of their insect prey, now and then darting down to the water, and skimming just above its glassy surface under the sweeping branches. The Hedge Sparrow's song sounds

Spotted Flycatcher arrives, 15th May.

Nightingales nesting, 7th May.

Wood Wren arrives, 20th April.

plaintively on every side, and the crescendo notes of the Wren are particularly sweet during the early weeks of May. In the still, warm evenings the Blackbird warbles, sometimes as he flies, or as he skips about among the bluebells and ragged robins in the marshy corners of the woods.

We must not let the spring go by without visiting some of the many beautiful nests which are built at this season. One of the first birds to cross a twig in nest-building is the sombre little Hedge Sparrow. Its beautiful home is often snugly placed amongst a heap of hedge clippings or in brushwood. It is made of green moss lined thickly with hair and wool and feathers, and the few twigs and roots round the outside lend it a rustic beauty peculiarly its own. In this nest the Hedge Sparrow lays five or six deep blue and spotless eggs. In a hole in some broken-down wall, or between the gnarled roots of a tree by the brook-side, or amongst the ivy on a bank, the Robin builds its bulky cradle of dead leaves, dry grass, moss, rootlets and hair, in which she lays her six white eggs, thickly marbled with brown. Another nest that we are sure to come across in the orchards or the woods and hedges is that of the Chaffinch. No bird architecture in Europe equals in beauty that of this gay little bird. The materials of the nest vary a great deal, but they are always beautifully arranged. Sometimes the walls of the nest are studded with silver lichens, at others bits of paper, spiders' webs, cocoons,

[Marginal notes: Hedge Sparrows laying, 5th April. Robins begin to lay, 10th April. Chaffinches begin to build, 5th April.]

and scraps of bark are used to assimilate the colour of the nest with surrounding objects, with a view to its concealment. The Blackbird's nest by the sides of the stream, or the Missel-thrush's high up in the forest trees, are also charming objects to be met with during a spring ramble through the woods and fields. Nor must we forget to give the Song Thrush a call as we pass her nest in the evergreens. Approach with caution, and you will see the mother bird sitting quietly on her eggs, her head peeping over one side of the nest, her tail over the other, and, as likely as not, her bill wide open. Perhaps she slips quietly off, or flutters away with harsh tumultuous cries, leaving her blue, warm eggs lying so temptingly in their hard cup, which is almost as smooth and round as if it had been turned on the potter's wheel.

Song Thrushes begin to build, 12th March; eggs, 20th.

On the banks of the stream, where the water leaps from rock to rock in mad career and whirls round the stones and by the deep banks, we may chance to meet with the domed nest of the Dipper, snugly placed among the overhanging roots. It is a strange and beautiful structure, made almost entirely of moss, which the spray keeps moist and green, and lined with dry grass and layers of dead leaves. The eggs are pure white, and four or five in number. The charming Gray Wagtail is another bird of these northern trout-streams, and regularly resorts to them in spring for the purpose of rearing its young. The

Dippers building, 7th April.

nest is generally made close to the water's edge, under a stone, or beneath the shelter of a tuft of fern or mass of bramble. Year after year the old birds return to a favourite nesting-place, and when their young brood is out there are few more charming sights than that presented by the parents' anxious and incessant care for their helpless little ones. A rookery in early spring is an interesting place to visit. From the gray dawn until the dusk of evening the Rooks are busy building their nests in the tall trees. Nothing can be done in a rookery without noisy consultation, and during the whole day an incessant chorus of cries is kept up. Many birds are simply repairing their old nests; others, not so fortunate, are building entirely new ones. Rooks are flying to and from the adjoining fields with roots and turf; others may be seen breaking twigs from the trees. Towards the middle of March the eggs are laid, but the busy scene loses none of its animation. All is bustle and life until the young are safely reared, and able to follow their parents to the pastures.

<small>Rooks begin building, 1st March; eggs laid, 10th March; young out of nests 18th, April.</small>

The woods are also made lively with birds. In the quiet recesses where man seldom treads, the shy Jays and Magpies are busy bringing up their families, using all their cunning to escape the vigilant gamekeepers. The latter bird is the first to breed, building its large nest high up in the branches of some forest giant, and often returning year by year to the old abode, patching it up, and increasing its bulk every season. Magpies often lay as many as eight eggs, and if the first clutch be destroyed another one is produced. The Jay loves to nest in the lower vegetation. Nothing suits it better than to build in some tall holly in the woods, or amongst the honeysuckle growing in a tangled mass over the nut bushes. Its nest is smaller than the Magpie's, has no roof, and is not used more than once. In the holes of the decaying timber the Titmice make their nests; but the Great Titmouse often builds a domed structure in the old nest of a Magpie, or even in the disused "drey" of a squirrel. Blue Titmice love the old elm trees, or a knot-hole in some sturdy oak. Coal Titmice, as a rule, breed nearer the ground, in an old decayed stump, or even in a cavity amongst roots. In the country lanes, and on the rough commons, as well as in the coppices, the Grasshopper Warbler reels off his monotonous music at intervals, whilst the loud, shrill *heel* of the Wryneck informs us that this singular species is back again for the summer. The peculiar habits and characteristics, the extraordinary fecundity, and the beautiful adaptation of structure to requirements, render the Wryneck specially interesting to the naturalist. In the larger woods the Carrion Crow still manages to exist, in spite of guns, traps, and poison. It is a late breeder—one of the latest of the Crow tribe —waiting until the leaves are sufficiently out to shield its domestic arrangements from the prying

Side notes: Magpies building, 29th March; laying, 10th April. Jays have eggs, 4th May. Carrion Crows laying, 29th April.

eyes of enemies. It returns to its old nest year after year if left unmolested. Another bird of the woods, though, alas! very locally distributed now, is the Hobby. It is a summer visitor to this country, and almost invariably selects the deserted nest of a Carrion Crow in which to deposit its eggs. Here and there, in highly favoured woods, the timid Woodcock breeds, making its slight nest among the withered bracken or in the drifts of dead leaves at the bottom of the tree trunks, in which the hen-bird deposits her four handsome eggs. The woods at eventide echo the harsh, grating crow of the Pheasant; and, in the fields adjoining, the Partridges, now in pairs, hold their carnival in the settling gloom. Higher up the hills, amongst the birch and larch woods, the Black Game engage in their curious courtship, and the Siskins in the pine trees are preparing for their broods. Their tiny nests are extremely difficult to find, and bear a somewhat close resemblance to the home of the Lesser Redpole, though not quite so neatly finished. The five eggs are bluish-green, spotted with dark-brown and violet-gray.

As the spring days get longer and warmer, bird after bird begins domestic duties, each species contriving to hatch its young at the time when the food on which they are reared is most abundant. Now we see the Swallows and the Martins busy at their old nests, or building new ones. Bit by bit they attach their

Hobby breeds, 24th May.

Woodcocks begin laying, 20th March.

Siskins leave lowlands, 25th March.

Swallows building, 1st May.

nests to the walls and beams and eaves, forming first a cup or shell of mud, which is warmly lined with hair and grass and feathers. These birds may often be seen in chase of some feather or straw which has been blown into the air; and during the hot, dry seasons, especially, it is surprising how far they will travel to get mud moist enough for building purposes. We have known House Martins fly nearly a mile to heaps of mortar, when the ponds, where they usually obtained their material, have been dried up by long-continued drought. How interesting and charming it is to watch the various ways of birds, and to chronicle the endless instances of their thought and intelligence! To us it is one of the greatest sources of our pleasure to discover such evidences of the presence of animal intelligence—of the little *mind* so busily at work in such tiny creatures, whose mental attributes more closely resemble our own than the past race of philosophers ever dared to dream. The Spotted Flycatchers may be observed in their old haunts, sitting on the bare branches and old stumps, from which they ever and anon sally out into the air to catch passing insects. This little bird, when insect life is not much on the move, sometimes flutters down among the grass and weeds to disturb the flies, and then catches them as they rise into the air. It may be known by its oft-repeated notes of *chee-tic, chee-tic-tic-tic.* The pretty Whinchats sit upon the tall hemlock

House Martins have eggs, 25th May.

Whinchats laying, 30th May.

and dock stems in the grass meadows, where they cunningly conceal their nest in the thickest herbage. Their eggs are very beautiful objects—a delicate turquoise-blue with a zone of very faint brown spots, usually round the larger end. The Tree Pipit is another very prominent bird of the fields, and may be seen soaring into the air from almost every tree top, singing as he goes; returning to the old perching place to rest a few moments, then fluttering up again. He loves to sit on the dead branches, and will return season after season to one particular tree. He also nests in the mowing grass, or sometimes in the pasture, often under a tree by the hedgerow. The nest is a simple structure made of dead grass, a flake or two of moss, and lined with hair. The eggs vary considerably in colour, but all in the same nest are uniform in tint as a rule. Few birds wander so little from home, and during the entire breeding season it will remain in one particular field until the young can fly. At this season wherever there are birds there is music. The blue dome of heaven, across which the big white masses of fleecy cloud, eloquent of fine weather and calm, slowly and solemnly drift, is resonant with the Skylark's song. From all parts of the upper air the little brown specks of birds are warbling against each other; far and near their melody is sounding, sweet as freedom's charm only can make it. Who, with a spark of love for Nature and her works, can find it in their hearts to

Tree Pipit arrives, 18th April.

Skylarks in fullest song, April and May.

cage a Skylark? An imprisoned chorister such as this is an insult and a blot upon the spring. Almost every species of bird in spring-time makes some attempt at song. The glad feelings of joy which the vernal year inspires in every living creature must be expressed in one way or another. Even such birds as Ravens and Magpies modulate their hoarse voices into much more musical cries; the House Sparrow's chirp has now more cadence; the Titmice string their notes together into a vernal song. Bird music is everywhere; and the crash of melody that comes with spring echoes and re-echoes from every corner of Nature's fair domain. *Magpies chatter, 1st to 31st March.*

Before leaving the early-breeding birds, we must not fail to visit a colony of Herons. We best associate this long-necked and long-legged bird with the water-side, standing as if lost in thought near the pool patiently waiting for his meal of frogs and fish; but if, when he unfolds his broad black wings and rises into the air, we follow him home, we shall find that he makes his nest in a tree, where he also generally roosts, and where he is as much at ease as any of the perching birds. Herons build in colonies like Rooks, using one particular spot from time immemorial, making their big flat nests of sticks, and turf, and roots, on the broad branches of the fir and other trees. The old birds may often be seen perched on the very topmost spike of a larch tree, and most ungainly objects they look as they *Herons begin repairing their nests, 25th March.*

crane their long necks from side to side and peer down through the foliage at the unwelcome intruder. The air is full of Herons sailing about so buoyantly, and here and there you may hear the big birds crashing through the branches, either hastily leaving their nests or coming back to them. Soon the pale green eggs are laid, and when the young are hatched the scene becomes even more animated and interesting. Regularly as Rooks the old birds pass to and fro with food, and the ground under the nest trees is strewn with broken egg-shells, dead nestlings, which have either fallen or been blown from the nests, and quantities of decaying fish. When they are partly fledged, and long before they can fly, the young birds climb out of the nests into the branches, using their beaks as well as their claws to assist them. Unfortunately the poor Heron is a much-persecuted bird, and its colonies are few and far between, and only in districts where man affords them his protection. It is cause for regret that " our last large bird " should bid fair to become extinct in England as a breeding species, and share the same fate as the Bustard, the Spoonbill, and the Kite, which have now ceased to rear their young in this country. Another early-breeding bird is the House Sparrow; in fact, it is busy rearing family after family during all the warm months of the year. Some of these birds build a domed nest in trees, others a slovenly structure in holes of buildings

Marginal note: Herons laying, 7th April.

or in dovecots. This double type of nest is exceedingly interesting, and is a question still unsolved by naturalists. Even the common Sparrows of the house-top invite careful observation, and their habits, stages of plumage, and economy generally are by no means perfectly known or understood.

The larger birds of prey are exceptionally early breeders, but the smaller species nest later when their food becomes more abundant. First of the smaller Hawks to breed is the warlike Sparrowhawk. He and his mate build a flat nest for themselves, sometimes high up in the forest trees, sometimes in the lower larches and firs. The eggs are remarkably handsome, pale bluish-green, spotted and blotched with rich brown and gray. The Sparrowhawk's fecundity is large, and it will go on laying egg after egg in the same nest for weeks together. This bold warrior bird plays sad havoc amongst the Willow Wrens and Goldcrests, both these species being his favourite food in early spring. A little later the Kestrel or Windhover commences breeding. This interesting little bird never makes a nest for itself, but takes possession of a deserted Magpie's or Crow's nest. Sometimes it lays its eggs in a hole in a cliff or ruin. The six eggs are laid on whatever material may chance to be there. The hen performs the greater part of the task of incubation, the cock keeping her regularly supplied with food. This pretty little Hawk may easily be recognised,

Sparrowhawks laying, 21st April.

Kestrels commence breeding, 25th April.

even at a distance, by its peculiar habit of suspending itself in the air on almost motionless wings, while it surveys the ground below. Few birds are more useful to man, its principal food consisting of mice and beetles.

The shrubberies in spring-time are the chosen haunt of many birds that love seclusion and concealment. Shy Bullfinches whisper love songs to their mates among the yew and holly trees; Greenfinches build their rustic nests among the evergreens; and many a shy Warbler delights to suspend its flimsy cradle in the briars and brambles by the stream which trickles through these secluded places. The pleasant notes of the Greenfinch at this time of year sound particularly sweet, especially when several males are warbling in company. It is a noteworthy fact that this bird is very social among its kind during the breeding season, and several nests are often built quite close together. We have frequently found two nests in the same bush, and once or twice have seen them almost side by side.

Bullfinches laying, 22nd April.

Greenfinches nesting, 24th April.

Let us now leave the woodlands and the fields and see what spring is like on the moors and mountains. On the mild, genial days in April we may stand and watch the Snipes careering high in air, and listen to the peculiar drumming sound which they make when engaged in courtship. The Jack Snipe has left us for much higher latitudes, but the common Snipe frequents some quiet bit of boggy ground either on or near

Snipes drum, 1st April.

Jack Snipes leave, 28th March.

the moor, laying four richly-marked, pear-shaped eggs in a scanty nest under the shelter of a little willow bush or tuft of rushes. On the breezy, open moors, the cock Red Grouse are crowing from almost every little hillock, their heads being visible above the tall heather, and as we walk along they rise from our feet on whirring wing, uttering their loud cries of *go-bac, go-bac, go bac-bac-bac* with startling distinctness. The Golden Plover finds a congenial home in the marshes on the rough tablelands at the mountain tops, and makes its slight nest on the ground, laying four very handsome eggs, something like those of the Lapwing, only yellower. Lower down the hillsides, on the commons and rough summer fallows, the Peewits rear their young. These birds are more or less gregarious in the breeding season, and numbers of their nests may be found quite close together. Amongst the big boulders of rocks on the moors the Ring Ousel, fresh from a southern haunt, sits and pipes his wild song, viewing us all the time with suspicious glances. He is something like the Blackbird in appearance, but may be readily distinguished by the broad crescent of white across the chest. He makes a very similar nest to that species, and the eggs cannot be told from those of the Blackbird, requiring careful identification. The Twite is another bird of the moor, and rises disturbed from the heather as we walk along the rugged sheep-

Snipes laying, 14th April.

Red Grouse nesting, 1st April to 31st May.

Golden Plover laying, 12th May.

Peewits have eggs, 7th April.

Ring Ousels nesting, 24th April.

Twite begins nesting, 10th May.

paths. It sits on the sprays of vegetation, and utters at intervals its long-drawn, plaintive note of *twite*. Even on the moor itself numbers of Peewits are nesting as well as Dunlins, and the eggs of the former bird are eagerly sought by the country people, who send them to the markets, where they are held in high esteem for the table. On every bit of marshy ground the Meadow Pipit abounds. This species comes from the lowland pastures to rear its young on the breezy moors, making its nest on the ground under a tuft of grass or rush, or beneath a bush of heath. The Meadow Pipit is very frequently the foster parent of the Cuckoo, those errant birds searching everywhere for the "Titlark's" nest, in which they drop their alien egg. In the gorse coverts, now glowing in the sunlight like masses of burnished gold, the sweet-songed Linnets make their nests. They, too, like the Meadow Pipit, leave the lowland pastures for the spring and summer. Another bird of the gorse coverts is the gay little Stonechat. Pairs of this bird may be seen flitting about the covers, the male often sitting on some bending spray of furze and warbling his simple song. The Merlins also are back again in their accustomed summer haunts waging relentless and incessant warfare on the small birds and weakly Grouse. This dashing little Falcon, the pluckiest of its race, returns unerringly to one particular spot, and nests year after year in one chosen place, this season's eggs being laid but a few

Marginal notes:
- Dunlins have eggs, 10th May.
- Cuckoos laying, 24th May to 30th June.
- Linnets breeding, 30th April.
- Stonechats have eggs, 28th April.
- Merlins laying, 13th May.

feet from where those of last were hatched. These upland solitudes are full of charms for the naturalist. They bring him face to face with Nature in her unchanged aspects. Endless incidents as strange as they are interesting, and countless objects as novel as they are beautiful, are constantly crowding themselves before his observation.

During the latter days of spring bird life on the coast undergoes many important changes. As a rule, however, the land birds are the first to feel the influence of the vernal season, water birds being later in almost every respect. Spring is nearly over ere the graceful Terns come back to their breeding stations; and very few shore birds have eggs until the beginning of the warmer months. Spring and summer time among the sea birds is full of equal interest. A naturalist's first visit to a colony of sea-fowl is a red letter day in his experience; and no matter how many times his visits may be repeated, there always exists the same charm and the deeply-absorbing interest of his first impressions. The writer has had the good fortune to visit most of the celebrated bird stations round the English and Scotch coasts, and his experiences of the stirring, vivid scenes are indelibly impressed on his memory. To see these bird colonies at their best, a visit should be paid in the late spring and early summer months. It matters little which locality is chosen; all are full of interest; but certain birds in many cases confine themselves to certain districts. For instance, nowhere can the nesting economy of Terns and Gulls be better studied than at the Ferne Islands; in no other locality in the British Islands but St. Kilda can the ornithologist watch the Fulmars at their nests in such enormous numbers; while the Bass Rock must ever be associated with the Gannet, and the cliffs of Flamboro' offer unusual facilities for observing Guillemots and Razorbills. On these latter rocks every available ledge is crowded with Guillemots, and the crannies in the cliffs are tenanted by countless Razorbills. Viewed from the sea these brave old cliffs present a wonderful sight. From all parts of the ledges birds are incessantly flying down to the sea; whilst the face of the rocks is almost hidden in places by the hordes of birds that are passing up and down. But the best place in this country to see the Guillemot at home is on the "Pinnacles," a group of stack-like rocks at the Ferne Islands. They are some sixty feet high, and the flat table-like tops are one moving mass of birds, most of them sitting on their solitary egg. When approached by human intruders, the whole mass of birds hurry off, pouring from the edges in streams, helter-skelter into the boiling sea below, leaving the surface of the rock thickly strewn with their beautiful eggs of almost every conceivable hue. In the hurry of departure many eggs get knocked off into the sea. On the ledges of these cliffs

Marginalia:
- Fulmars lay, 15th May.
- Guillemots and Razorbills laying, 15th May.

many Kittiwakes build their nests, and their noisy clamour sounds high above the roar of the sea. On the flat islands adjoining these "stacks," vast colonies of Gulls breed, each islet, for the most part, being monopolised by a single predominating species. One island is tenanted by Terns, another by Herring Gulls and Puffins, a third by Lesser Black-backed Gulls, a few Eider Ducks and Oystercatchers, whilst another outlying rock is tenanted exclusively by Cormorants. As visitors approach, all becomes uproar and tumult, and the air is soon full of fluttering birds. The scene can only be compared to a heavy snow-storm. Under foot their eggs lie so thickly that great care must be exercised in walking about, or we should inevitably trample upon them; and in every available place amongst the sea-campion and coarse grass we stumble across a nest. The island where the Puffins breed is undermined with their burrows, and the ground is literally full of birds, each sitting on its solitary egg.

The Bass Rock, with its colony of Gannets, is equally full of interest, the nests being crowded thickly on every bit of cliff and rock ledge at all suitable to hold them. Early in spring the old Gannets return to the rock, and in April and May the scene is animated beyond all description. For some considerable distance round the rock the Gannets may be observed fishing. They catch their finny prey by rising high in air, and then with closed wings darting headlong down

(margin notes: Kittiwakes laying, 18th May. G la May. Puffins laying, 10th May.)

D

into the water with a terrific plunge, rarely failing in their efforts. But if the air is full of birds the rocks are even more so. You may see them standing on the short grass at the top of the cliffs fast asleep, with their heads buried in their dorsal plumage. Thousands of birds are flying to and fro along the face of the lofty cliffs; on every possible place a Gannet's nest is built, and the birds are incessantly quarrelling, owing to their close companionship. The air resounds with their loud cries of *carra—carra—carra*; and as we approach the nests, numbers of birds rise from them, and, after disgorging a half-digested fish, launch buoyantly into space. Some birds refuse to leave their egg until they are pushed from it, all the time uttering cries of angry remonstrance. Occasionally two birds fight fiercely, and go toppling over the cliffs down the dizzy depths locked in tight embrace, but separating as they fall, and recovering themselves long before they reach the water. Many of the Gannets in the air are carrying materials for their nests. The Gannet only lays one egg, which is nearly white when fresh, but soon becomes coated with dirt from the wet feet of the parent bird. The nests are built of seaweed, coarse grass, and turf, with a few feathers by way of lining; and the stench arising from the decaying fish and slime, and the droppings of the birds, which make the rocks look as if they had been whitewashed, is sometimes almost unbearable. On the stretches of shingly

Gannets begin laying, 7th May.

beach, and in the quiet bays just above the line of drift which marks high tide, the Oystercatcher lays its eggs. This singular bird often arranges several nests among the shingle before selecting one for its purpose; the female lays three or four eggs—buff in ground colour, spotted and streaked with blackish-brown and pale gray. The old birds become very anxious for their speckled treasures when man appears upon the scene, and rise screaming in the air, hurrying to and fro, piping mournfully all the time. Another well-known bird of the coast in spring-time is the Ringed Plover. This species breeds upon the stretches of fine sand, eschewing the coarser shingle. This is because the colours of its eggs harmonise best with the sand, being buff in ground colour, finely spotted with blackish-brown and gray. Another bird which often breeds on the moorlands near the sea, as well as in the inland districts, is the Wild Duck. This bird breeds early; and as soon as the eggs are laid, the drakes desert the ducks, and leave them to bring up their broods by themselves. Sometimes the Wild Duck will make its nest a mile or more from the water.

Oystercatcher have eggs, 10th May.

Ringed Plovers have eggs, 24th April.

Wild Ducks laying, 7th April.

The most prominent features of bird life in spring-time are love and song, marriage and family cares. The songs of birds are a most fascinating subject for the naturalist's inquiry. To me it is always a never-failing source of interest to observe the gradual increase of bird-music through the spring-time until it reaches perfection, and then as imperceptibly dies away again, like the swelling volume of melody from some stupendous organ-pipes, commencing in low and fitful strains, then rolling out in grand, majestic cadence, then fading away again as gradually as it rose. The following table may therefore prove of interest to the student of natural phenomena, and serve as a guide to more elaborate observations. It has been compiled from my note-books, extending over a long series of years, and some nineteen species of birds have been selected as examples:

Name of Species.	March, 31 days. Sang on	April, 30 days. Sang on	May, 31 days. Sang on
1. Missel-thrush	20 days	First 16 days	Silent
2. Song Thrush	31 days	30 days	31 days
3. Blackbird	16 days	30 days	31 days
4. Robin	31 days	30 days	31 days
5. Wren	10 days irregularly	30 days	31 days
6. Hedge Sparrow	31 days	30 days	31 days
7. Willow Warbler	Absent	Last 20 days	31 days
8. Chiffchaff	Absent	30 days	31 days
9. Whitethroat	Absent	Last 6 days	31 days
10. Blackcap	Absent	Last 20 days	31 days
11. Redstart	Absent	Last 14 days	31 days
12. Starling	31 days	30 days	31 days
13. Meadow Pipit	Silent	17 days irregularly	31 days
14. Tree Pipit	Absent	Last 5 days	31 days
15. Chaffinch	27 days	30 days	31 days
16. Yel'o v Bunting	23 days	30 days	31 days
17. Greenfinch	Last 4 days	30 days	31 days
18. Skylark	20 days	30 days	31 days
19. Cuckoo	Absent	Last 8 days	31 days

Towards the close of spring the young of many of the early-breeding birds are out of the nests and well able to fly. In many cases these species rear a second and even a third brood during the course of the season. Many young Thrushes are abroad by the end of April, also Robins, Hedge Sparrows, and Rooks. The mortality amongst these young and inexperienced birds is vast. Of the tens of thousands which leave the nests, but a very small percentage escape their numerous enemies and reach the adult stage of existence. We may meet with these nestlings everywhere, fluttering among the bushes or hopping along the ground, too feeble to make a prolonged flight, and offering a tempting bait to the crafty Crows and Hawks, and the wandering weasels and rats. But bird life is prolific, and brood after brood appears to fill the gaps which are constantly being made in the endless struggle for existence. Spring-time, however, is now nearly over. Bird life at this season is so active, and so much requires our observation and attention, that it is difficult to make a selection from the rich store before us. All we can do is briefly to allude to a few of its most prominent incidents and most interesting phases, and leave the short and imperfect sketch to be further elaborated by the observer himself.

CHAPTER III.

STRANGERS OF THE SPRING.

IN scanning the list of British birds, the naturalist will note with regret the names of many species now marked as accidental visitors only which were once regular migrants to our shores. Modern improvements and ruthless persecution have exterminated these birds as breeding species; yet scarcely a spring passes that some of them do not wander across the Channel, prompted to do so perhaps by memories of more prosperous days, or following a deeply-rooted desire to visit English haunts, and breed in districts where, unmolested, their ancestors had bred for so many generations in the past. Many of our accidental visitors only reach the British Islands in spring. Their breeding range does not extend so far north as this country, so that they do not come within the influence of the southern stream of migrants in autumn; they reach our shores by travelling too far north when on their way in spring to their summer quarters. They are, for the most part, birds that migrate north at that season, either from their winter home in Southern Europe or Africa. Some cross the English Channel as a matter of course, but are rarely allowed to live here in peace, being hunted to death by "collectors"; others overshoot the mark, and doubtless cross to us with the usual stream of spring migrants. The great migration of birds in spring is from south to north. So far as we can ascertain, there is little or no migration from the east at this season, and our feathered strangers of the spring invariably come from the south. The gravest of doubt attaches to the *bona fides* of any eastern bird said to have been shot in spring, as to any southern bird obtained in autumn. Students of British birds should always bear this golden rule in mind.

When we come to treat of strangers of the autumn, we will discuss more fully the movements of these wanderers and the routes they travel; for the present we will confine our attention to a brief notice of the various interesting species that make their appearance here in spring. The Hobby and the Honey Buzzard may be dismissed with little notice, inasmuch that both of these birds are regular if rare migrants here in spring, and breed very sparingly in one or two favoured localities. Precisely the same remarks apply to the Hen Harrier and Montagu's Harrier. Both these birds were once much more common and widely dispersed than they are at the present time; nevertheless, a few scattered pairs come here to breed every summer. All these birds of prey are regular migrants to Northern and Central Europe, and winter in the basin of the Mediterranean, or in Africa, as far south as the Cape. The Black Kite is a regular summer visitor to the forests of Central Europe, and the example which was obtained on our shores doubtless lost its way in coming north in spring. Very similar to the preceding species, in its summer distribution in Europe, is the Scops Owl, which, from time to time, pays us an uncertain visit in the spring; yet, so far as is known, neither of these birds has ever bred in the British Islands—they are purely accidental wanderers from their usual haunts.

The Rock Thrush is another straggler here in spring, far out of its usual habitat, its nearest breeding-place being the Hartz Mountains, in about the same latitude as London, though 500 miles to the east of that locality. Nearly all the rare Warblers that pay us their irregular visits are strangers of the spring. The only wonder is that these species have not regularly extended their migration to this country; but the English Channel appears to be a great barrier to many migratory birds. This is one of the many peculiarities of geographical distribution. In a great many cases we find narrow seas forming the line of demarcation between widely different faunas; and so far as the British Islands are concerned, this seems to show that the birds which breed as near to us as France and Holland, and yet do not cross to this country, have only extended their range so far north after Great Britain became separated from continental Europe, our regular migrants probably acquiring the habit of breeding in them before they became islands at all. Birds, we know, are remarkably conservative in their habits, and would continue to visit their British breeding-grounds, undeterred by the narrow channel that eventually separated these islands from the main land. Of the five species of Warbler that stray here in the spring, three of them, the Aquatic Warbler, the Great Reed Warbler, and the Icterine Warbler are regular summer visitors to France, and breed there in more or less abundance. The Barred Warbler actually passes the latitude of the British Islands, and breeds as far north as the south of Sweden; and this species should be watched for, especially in autumn, as it is highly probable that it has often been overlooked in this country. The Orphean Warbler, although it has been said to have bred in our islands, can only be looked upon as a straggler to them, not breeding regularly nearer than Luxemburg. The Golden Oriole is another stranger of the spring, yet one to which the English Channel is almost an impassable barrier. Although it is common enough in all parts of Central and Southern Europe suited to its requirements, it only occasionally crosses over to us. In the same way the Woodchat Shrike very rarely extends its migrations to our islands, although it breeds in the south of France, in Holland, and in Germany. The White Wagtail and the Blue-headed Wagtail are also common enough in the meadows and on the banks of the streams on the continental side of the Straits of Dover, yet only accidentally cross over to England. These two species, however, are specially interesting to British naturalists, for it would seem that they are gradually overcoming their continental prejudices, and occasionally rear their young in this country. Both should be eagerly looked for by British field-naturalists.

A far different spring-tide stranger is the Alpine Swift, which pays us uncertain visits on its migrations to the mountain districts of Central and Southern Europe, where it rears its young. This fine bird is an inhabitant of the mountains, delighting in the ravines and passes, and showing more preference for the rocky wilderness than for the habitations of man. The occurrence of the Isabelline Nightjar (an eastern species) in England during the early summer, and of the Red-necked Nightjar (a southern species) in autumn, are events of exceptional interest to naturalists, and at first sight seem very serious exceptions to those laws which appear to govern the seasonal movements of birds. Had the dates been exactly reversed, there would have been nothing extraordinary in the appearance of these Nightjars. The most feasible explanation appears to be that the eastern Isabelline Nightjar got attached to a stream of migrants starting northwest from that bird's winter quarters somewhere in the Soudan, and instead of reaching its usual habitat in Turkestan, it strayed into Western Europe. The southern Red-necked Nightjar undoubtedly came north in spring, probably with a party of Common Nightjars, from West Africa, and managed to escape detection in this country until it was shot in the autumn. By a singular coincidence another example of the Isabelline Nightjar was shot on Heligoland one day earlier in the same month, but eight years previously. Another has been obtained in Sicily, and others in Malta.

Flocks of the beautiful Bee Eater have from time to time appeared in this country in spring, individuals which have overshot their mark on their vernal migration to Central Europe from South Africa. The Hoopoe is another of the birds which the British ornithologist looks upon with regret, for there can be no doubt that this singular and handsome species would become a regular summer visitor to our shores, and rear its young upon them, if it were protected and encouraged. It has come almost every season, and has bred here from time to time. The Great Spotted Cuckoo has been known to visit the British Islands, having inadvertently strayed north of its usual summer quarters in the Spanish peninsula. No less than seven species of the

Heron tribe are accidental wanderers to Britain in spring: birds which have misjudged their distance, or have been driven out of their usual course during migration to their breeding-quarters in Central Europe. Of these, the Great White Egret, the Little Egret, the Squacco Heron, the Buff-backed Heron, and the Night Heron, are purely wanderers from their usual habitat; but the Little Bittern and the Spoonbill were once birds of regular passage to this country, and used to breed in our marshes and fens before they were drained and "improved" away. Fancy a colony of Spoonbills in Norfolk now! In like manner the magnificent Crane was a regular frequenter of our marshes, and even now from time to time pays a passing visit to the scenes of its ancient home—an outcast and an exile; but the delicate Demoiselle Crane is only an accidental wanderer here, doubtless from the wide-extending lagoons of Spain.

A sad interest is also attached to the fitful visits of the Avocet. Formerly this beautiful bird bred in the marshes of the low-lying counties; but drainage has destroyed its old familiar haunts, and nothing but persecution awaits the few stragglers that are tempted to journey here by memories of the old and prosperous summers, when they had the good fortune to be allowed to rear their broods in peace. The Black-winged Stilt is only an accidental wanderer from the Spanish marshes, and at no time ever appeared to frequent our islands for the purpose of reproduction. It should be remarked that spring is the usual season for examples of rare Terns to visit the British Islands—birds which wander here on their migration northwards; just as autumn is the time to expect stranger Gulls, which are driven from their home in the Arctic regions by severe weather. The Black Tern still visits the lowland broads in spring, where it formerly used to breed; but the examples of the White-winged Black Tern and the Whiskered Tern that reach this country are lost and wandering birds, out of their usual range. Of the Petrel family, the Dusky Shearwater is an occasional wanderer to the British seas in spring from its island homes off the west coast of Africa; and Wilson's Petrel, a southern hemisphere species, though a stranger of the spring to us, is in reality a winter wanderer from its haunts in lonely Kerguelen Island in the remote Indian Ocean. Our last stranger of the spring is the Garganey. Many individuals of this species, on their way north from their winter quarters in the basin of the Mediterranean, visit the British Islands, and some of them remain to breed in districts where they are left unmolested. This bird was much more plentiful with us before the larger fens were drained.

It is worthy of remark that all, or nearly all, the birds that accidentally reach us in spring are late migrants to Europe, and come with the last wave of migration that spreads northwards at this season. On the other hand, such species as the Hoopoe, the Avocet, and the Spoonbill, which were once regular visitors to our shores, do not migrate exceptionally late. These facts seem to confirm the theory that those species, individuals of which are purely accidental visitors in spring, have only extended their range into latitudes as far north as the United Kingdom since our islands were detached from the Continent. They were the last birds to push north again after the glacial ice receded, and were evidently disinclined to cross the waters of the English Channel, which to them was a strange and unknown sea. All Arctic birds are the latest of migrants, tarrying in their winter quarters long after birds of temperate climes have hastened away, waiting for the summer, which spreads rapidly over their far northern haunts with but few warning signs of spring. The Great Reed Warbler and the Barred Warbler do not visit us, not because there are no haunts in this country suited to their requirements, but because they have never learnt the way across the Channel which divides us now from the Continent; the Reed Warbler and the Blackcap do visit us because it has been their custom to do so ever since the glacial period passed away, and when this country was probably a peninsula of Western Europe. The following table of specific characters will enable the reader to identify any of these strangers of the spring in their breeding plumage:

Species.	Points of Distinction.
Hobby	Similar to Peregrine, but smaller; breast striped not banded. Thighs and vent chestnut. Length 12 to 14 inches.
Honey Buzzard	Distinguished from all other Raptores by having space between eye and bill covered with feathers. Length 21 to 25 inches.
Hen Harrier	Outer web of fifth primary notched; tail barred. Length 22 inches.
Montagu's Harrier	Outer web of fifth primary entire; tail barred. Length 18 to 19 inches.
Black Kite	Tail uniform brown, fork less than 2 inches deep. Length 22 inches.
Scops Owl	No operculum; toes bare of bristles and feathers. Length 8 to 10 inches.
Rock Thrush	Tail chestnut; head and neck of male blue; female mottled brown, but tail red as in male. Length 7½ inches.
Aquatic Warbler	Similar to Sedge Warbler, but yellower, with two dark bands along the crown, and pale line in centre. Length 5 inches.
Great Reed Warbler	Similar to Reed Warbler in general colouration, but centre of belly nearly white; tail graduated. Length 8 inches.
Icterine Warbler	Similar to Wood Wren on upper parts, but trifle more olive, underparts greenish yellow. Length 5¼ inches.
Barred Warbler	Upper parts brownish gray; under parts grayish white, barred with

Species.	Points of Distinction.
Barred Warbler (*continued*)	dark brown, as are also the lower back, rump, and upper tail coverts. Birds of year have under surface barred only on under tail coverts. Length 6¾ inches.
Orphean Warbler . .	Similar to Blackcap, but outer web of outermost tail feather on each side white. Length 6¾ inches.
Golden Oriole . .	Body yellow generally; wings black, tail black and yellow. Female and young much less brilliant. Length 9 inches.
Woodchat Shrike . .	Forehead black; crown, nape, and upper part of back chestnut; lower back black, rump white; white alar bar. Female duller: young brown above, buff below, mottled and barred with rufous brown and dark brown. Length 7½ inches.
White Wagtail . .	Similar to Pied Wagtail, but back gray. Length 7¼ inches.
Blue-headed Wagtail .	Body yellow generally, lightest below; throat and eye-stripe white; ear coverts gray. Female duller. Length 6½ inches.
Alpine Swift . . .	Blackish brown above; under parts white, except brown chest band. Length 8¼ inches.
Isabelline Nightjar . .	General colour mottled buff and gray; bases and outer part of inner webs of wing feathers white; white spots on wings and tail absent. Length 10½ inches.
Red-necked Nightjar .	Similar to Common Nightjar, but larger; nape rufous. Length

Species.	Points of Distinction.
Red-necked Nightjar *(continued)*	13 inches (against 10½ in Common Nightjar).
Bee Eater	Forehead white, shading through bluish into chestnut on crown; throat yellow bordered by narrow black line; remainder of under parts green; tail green, two centre feathers an inch longer than rest. Female slightly duller. Length 10½ inches.
Hoopoe	General body colour buff and white; conspicuous crest rich buff tipped with black, hindermost feathers with white and black; tail black with broad bar of white. Length 11 inches.
Great Spotted Cuckoo	Head gray and crested; upper parts brown; under parts white. Length 15½ inches.
Great White Egret	Uniform white; dorsal plumes assumed in pairing season. Length 36 to 43 inches.
Little Egret	Same as preceding in colour, but in pairing season has several long pendent feathers from crown in addition to dorsal plumes. Length 20 inches.
Squacco Heron	Similar to preceding in colour, but head, nape, neck, back, upper breast, and dorsal plumes are buff, the feathers on the head banded with black; elongated pendent feathers from crown white margined with black. Length 18 inches.
Buff-backed Heron	General plumage white; head,

E

Species.	Points of Distinction.
Buff-backed Heron (*continued*)	neck, and breast feathers filamented, and buff in colour; dorsal plumes buff. Length 18 inches.
Night Heron	General colour lavender; head and upper back black shot with green; crest composed of several long white cylindrical feathers. Young brown above, spotted with white; white below, striped with brown. Length 21 to 24 inches.
Little Bittern	Upper parts black shot with green; under parts buff; lower neck feathers elongated. Length 12 inches.
Spoonbill	General colour white; crest short and bushy, tinged with yellow; bill broadly flattened at end. Young have wing feathers tipped with black, and black shafts. Length 31 to 36 inches.
Crane	General colour gray; wings black; innermost secondaries forming long curled plumes; crown scarlet and warty, bare of feathers. Length 48 inches.
Demoiselle Crane	Similar in colour to preceding, but smaller; throat feathers elongated; tuft of white feathers on each side of head; secondaries forming uncurled plumes. Length 31 inches.
Avocet	Bill long, slender, and recurved; general colours black and white; feet semi-webbed. Length 18 inches.
Black-winged Stilt	Bill straight and slender; general

Species.	Points of Distinction.
Black-winged Stilt (*continued*)	colours black and white; legs about 10 inches in length and very slender. Length 13 inches.
Black Tern . . .	Head, neck, and under parts (except vent and under tail coverts, which are white) black, remainder dark gray; bill black; legs and feet dark brown. Length 10 inches.
White-winged Black Tern	Similar to preceding, but white shoulders; tail white; under wing coverts black; bill dark red; legs and feet scarlet. Length 10 inches.
Whiskered Tern . .	Top of head black, remainder white; under wing coverts white; belly black; bill and feet as in preceding. Length 11½ inches.
Dusky Shearwater . .	Smaller than Manx Shearwater; under parts, including lores, ear coverts, and axillaries white. Length 11 inches. (Length of wing, Manx Shearwater 9½ inches, Dusky Shearwater 8 inches.)
Wilson's Petrel . .	Similar in colour to Fork-tailed Petrel; tail square; tarsus covered with large plates or scales; feet black, centre of webs yellow. Length 7½ inches.
Garganey . . .	Upper parts similar in colour to those of Shoveller; shoulders blue; breast brown banded with black; under tail coverts white, spotted with dark brown. Female similar to female Teal, but green speculum absent. Length 16 inches.

E 2

CHAPTER IV.

BIRD ORNAMENTS AND TOURNAMENTS.

No other season of the year is so appropriate as spring for studying the various nuptial ornaments of birds, and the strange antics and battles the birds themselves engage in during the days of their courtship. The researches of ornithologists during recent years have shown the inseparable affinity between Beauty and Utility; the two subjects are practically synonymous. According to the old philosophy, Beauty existed for its own sake; it had no ulterior object or use beyond increasing the grandeur of the universe and gratifying the eye of man. But Beauty has not been acquired or developed aimlessly or in vain. Nature is far too conservative to permit such lavish waste; and we may rest assured that every spot on the wings of the butterfly, every spangle on the Humming Bird, every refulgent plume of bird, and gorgeous tint of shell and flower, have been developed for a purpose. So far as birds are concerned, this beauty of plume and feathered ornament is correlated with the function of reproduction to a very singular and extraordinary degree.

British birds, with one or two exceptions, are not remarkable for any great brilliancy of plumage; it is among the feathered tribes of tropical regions that colour attains its greatest variety and splendour. The reason for this is probably because gay birds would be exposed to enemies during a northern winter, when most of the vegetation is denuded of foliage; in the tropics, most of the trees and shrubs are evergreen and very dense in character, thus affording concealment to gaily-attired birds, which, in the great majority of instances, love to hide their brilliant dress in the deepest solitudes. Nevertheless, we have ample material to illustrate some of the uses of colour in the birds of our own woods and fields. This brilliancy of colour is almost universally confined to the cock birds; it obtains to greatest splendour on the dermal covering of the adult males, and serves as an attraction to the much more soberly-arrayed females. As the pairing season of the various species approaches, the male birds' plumage reaches its greatest brilliancy of colour; plumes and feather appendages are assumed; bare patches of skin, combs, and wattles deepen in hue. The males of the Pheasant and the Red Grouse exhibit this latter peculiarity, their combs and ear wattles assuming a much more brilliant scarlet in the pairing season. Cormorants, at the same period of the year, don the filmy neck filaments and white tufts on the thighs; just before the season of courtship, the Ducks and Divers don their wedding garments; and the various species of Herons and Egrets put on their gorgets, plumes, and crests to enter the lists in the tournament of love. The Plovers and the Knots and the Grebes also increase in beauty early in the spring-time, and the curious Ruff dons his collar-shield of feathers. Many of the Finches, too, become more brilliant, not by a change of dress, or assumption of new plumage, but by the abrasion of the pale, deciduary margins to the feathers which all the winter have concealed the more showy parts of the plumage; as, for instance, in the Greenfinch, the Chaffinch, and the Brambling. There can be no doubt that these pale tips are a great protection to the birds in shielding them from enemies all the winter, for, by the time they are abraded, vegetation affords more shelter, and the charming dress is ready in all its splendour just when courtship begins. In some of the Gulls and Terns, a delicate rosy tint suffuses the breast just prior to the breeding-time. In many species where the feathers are not absolutely changed before the pairing season, they increase in brilliancy; as, for instance, the scarlet on the breast of the Linnet, the glossy sheen on the plumage of the Starling, the Rook, the Carrion Crow, and the Magpie.

Thus we have now seen how so many birds look their best just previous to mating; now let us analyse the subject a little farther, and learn the cause and the use of this important proceeding. All the male birds that develop these various beautiful ornaments are excessively careful to display them in their courtship, and strive to attract and fascinate the duller-plumaged females by the sedulous exhibition of their charms. Of all the lower animals, Birds approach most closely to man in their æsthetic tastes and their love for the beautiful. They show off their charms and parade their graces with as much care as a Hyde Park dandy, and for a precisely similar purpose—the conquest of the gentle sex. We have only to watch the pairing antics of birds to be thoroughly convinced of this. By spreading their wings and tail, and puffing out their gaudy plumage, and parading to and fro, the Pheasants and the Grouse seek to win their fastidious mates; the Herons and the Egrets erect their crests and plumes and gorgets, and show them off to best advantage; the Ducks and other water-birds swim round and round the females with bobbing heads, displaying their wedding finery; whilst the Ruff erects his mottled collar, and, with every feather bristling, struts about the pairing place, exhibiting his charms. The smaller birds are equally careful to appear to best advantage. The Chaffinch and the Linnet distend their rosy breasts and expand their wings, showing out the white spots on the coverts and the pale margins of the quills. The House Sparrow, with wings trailing on the ground and tail spread out like a fan, struts round and round his mate; whilst the Goldfinch, whose greatest charms are on his wings, turns his body from side to side, and flashes his yellow-margined pinions in the light. Other birds, such as the Blackcap and the Reed Bunting, erect the most conspicuous feathers on the top of the head during courtship; whilst the Pigeons swell out their metallic loricated breasts, and display the iridescent hues with the greatest adroitness and skill. I have seen the Magpie expand his ample tail almost over his back, and spread out his black-and-white wings to their full extent, drooping them to the ground, and in a crouching attitude exhibit the beauty of his plumage to the very utmost before several evidently appreciative females. Males and females of this species regularly congregate in early spring, in districts where they are at all numerous, the cocks to thus display their charms and attract the hens.

Birds that are not gifted with any very remarkable beauty or brilliancy of plumage appeal to the female's fastidious tastes through their song. Nature, ever conservative and sparing, rarely endows a bird with any great musical powers in addition to fine plumes or gorgeously-tinted feathers. The dull-plumaged birds find their ornament and attraction in their song. All our finest songsters are remarkably dull and sombre in colour. Witness the Nightingale, the Willow Wren, the Song Thrush, and the Blackbird; yet these, and all other musical dull-plumaged birds, strive their utmost to out-sing their rivals during the period of courtship. These little singing-birds may be watched in spring-time warbling against each other, and working themselves up into a perfect state of frenzy by their voice, the melodious contest very often leading to a pitched battle for the favours of the females. The Willow Wrens will sing and sing at each other in the birch trees, then dart, like little meteors, through the branches, fighting and warbling almost at the same time. The Yellow Bunting will do the same; and Robins are so pugnacious at this season that each male will endeavour to out-sing all rivals within hearing, and follow up his triumphs by fighting long and fiercely for supremacy. The vocal music of birds, however, is not the only sound they are capable of making to further their courtship. Some birds, in paying their attentions to the females, or in the presence of rivals, rattle their quills together; other birds make various sounds by rapidly vibrating the wings and tail. In some species, certain feathers in the wings and tail have been modified in structure and form, so that when the birds are in the air, they can produce a variety of sounds, as, for instance, the "drumming" made by the Snipe. Thus, then, we see that the various feathered ornaments of birds, their gay plumes, their sweet, melodious songs, their strange sounds and singular antics, have been developed and acquired for a purpose

—to charm and attract the opposite sex; and to this process of sexual selection, and to this rivalry, must be ascribed a very considerable amount of the beauty and the singularity in the dermal covering and appendages of birds, and probably of all the melody they utter. There is much evidence to show that female birds appreciate brilliancy, and variety of colour, and musical attainments in the opposite sex, and also evince no small amount of caprice and fastidiousness in the choice of a mate.

We will now pass on to the second division of our subject, and glance at a few of the curious tournaments which many birds indulge in during the season of courtship. These pitched battles are most frequent and most deliberate among the Gallinaceous family of birds—species which are armed with special weapons of offence. One of the most interesting tournaments of this description to be seen in this country is in the haunts of the Blackcock. The observer must visit their pairing-grounds at dawn, and conceal himself amongst the surrounding vegetation. These are recognised meeting-places in the open parts of the woods which all the birds in the vicinity frequent to carry on their singular courtship. Soon the cock-birds will make their appearance, and go through a series of strange antics before the females. Battle after battle is fought and won, the rival males keeping up the contest until all the victorious ones have paired. So fiercely are these battles fought that, during the conflict, the combatants are utterly oblivious to danger, and may be approached and shot, or even taken with the hand. In some birds, as, for instance, the Spur-winged Plover, the wings are armed with long, sharp spurs; and even in our own Peewit the male has a tubercle on the shoulder which enlarges during spring, and which most probably assists him in his combats with rivals. Another most remarkable bird-tournament is held by the Ruffs in spring. Certain recognised meeting-places soon become bare of turf, owing to the numerous combats upon them. Though spurless, the Ruff fights fiercely with his sharp beak and powerful wings, the singular ruff of feathers round the neck being opened out as a shield to protect the body from harm; and the face is covered with warty excrescences, which doubtless serve as armour against the blows from an opponent's beak. It would seem that these tournaments are carried on in an orderly way, with much method, and subject to certain regulations, the young males never being allowed to show off their charms until they have fought their way to fame, and literally "won their spurs."

There are combats between most birds during the season of love. The display of charms, and the utterance of song, inevitably lead to combat; they are the incentives to the strife, and never fail to work upon the passions of the rivals until they reach such a climax that deadly combat is almost invariably the result. Rooks will fight in the early spring with such fierceness as to kill each other; I have known Robins do the same. Even the proverbially gentle Dove is pugnacious in the pairing season, and the meeting of rival males at certain times is sure to end in battle. Sparrows are exceptionally pugnacious, and though the fights seldom result in more than the loss of a few feathers, they are carried on with most amusing vigour and fierceness. The antics and grotesque attitudes displayed and assumed by many birds, either in their fighting arenas, or before the females in other places, is very interesting. Some birds strut about in all the pomp and pride of their brilliant dress; others gambol in the air, or run to and fro puffed out with amorous excitement; many erect their crests, and distend their brilliantly-coloured wattles to the best advantage—yet all have the same great end in view, the attraction of the opposite sex.

The various ornaments of birds and their powers of song certainly assume a greater and a higher interest when we associate them with the well-being of their possessors. Vital, indeed, is the importance of these characteristics; beauty of structure and harmony of sound have each been evolved on purely utilitarian principles. New and ever-increasing interest, therefore, attaches to the love and courtship of the feathered tribes; for verily these actions have a most important bearing upon some of the highest questions affecting organic life.

CHAPTER V.

SPRING-TIME ON THE MOUNTAINS.

ALTHOUGH the glories of the spring are slow to creep over the mountains from the lower valleys, there is much of interest taking place upon the hills even in the very earliest days of March. Winter has not yet surrendered his grasp upon these upland wilds before many species of birds, whose haunts are amongst them, have actually commenced the great duties of the year. With a forethought and a knowledge of the future which we cannot help admiring, these mountain birds, especially the various species of Raptores and the predatory Crows and Ravens, so contrive to hatch their young at a time when food for them is becoming plentiful. The inexperienced naturalist is apt to think that there can be little of interest among the mountain-tops until the very latest days of spring; but let us stroll thither this wild March morning, and take a chapter from their natural history.

Over the fields in the valley one or two Skylarks are soaring skywards, tempted into song by the fitful sunbeams, and a solitary Wheatear, more venturesome than the rest of his kind, has made his appearance among the peat-stacks. Let us hope he will not have cause to regret his haste; but the early spring days here in these northern valleys are often severe, and snow-storms and tempests threaten to exterminate even the most robust of the feathered tribes. Along the banks of the burn, which rushes down in a swollen torrent charged with the melted snow from the hills, the Dipper flits uneasily before us. He is a resident here, the water being always open, and a food supply ever at hand. He dives into the ice-cold water just below the falls, and disappears for several moments, rising again among the foam-flecks, and swimming gracefully to land. It is too early for the Siskins, and the Twite has not yet dared to venture from the lowlands; but a flock of Snow Buntings rise and circle in the air above a grass meadow which has recently been spread with manure, and a pair of Hooded Crows, impudent and cunning as is ever their wont, just keep a respectful distance out of what they doubtless consider harm's way. As soon as we get clear of the coppices and the fringe of birch and alder trees on the banks of the burn, the cold wind seems to pierce right through us, and the temperature sensibly lowers as we mount higher up the hillsides. The Red Grouse scurry from out the brown heather, and skim across the moors, *go-bac-bac-bac*-ing as they fly, or timidly watch our movements as they stand upon the boulders. They are just about to pair, and the red combs of the cock-birds glisten again in the sun. These hills are a favourite haunt of the Golden Eagles; they have their eyrie on the giddy sides of yonder frowning glen. Let us shelter from the cutting wind behind these friendly rocks, and, knee-deep in the withered heather, wait and watch for his feathered majesty's appearance. Eagles are confirmed wanderers, and may not come this way for hours, even at all; still, the Red Grouse will amuse us, and the Hooded Crows serve to pass away the time.

Fortune, however, favours us this morning. The pair of Eagles are at home, and the male bird soars proudly over the distant ridge, which separates us from the adjoining valley, and, after circling round and round high up in the clear, bracing air, swoops down into the rocky fastness, where, for years upon years, they and their ancestors have made their nest. Great splashes of droppings, which show out white and conspicuous on the rocks, mark out the exact site of the eyrie; but as it is made in a part of the cliff which overhangs considerably, and we should require enough tackle to storm a rampart to get into it, we must be content with viewing it from the broken ground on the opposite side of the glen. The eggs, however, are already laid, and the male bird regularly supplies his sitting partner with food. As soon as the young Eagles are hatched, the ewes will be dropping their lambs on the hillsides, and the old birds will .levy a costly tribute from them. Lamb after lamb will disappear, in spite of all the shepherd's care and watchfulness; but fortunately the gentleman-farmer is proud of his Eagles (a most unusual fact), shrugs his shoulders at their depredations, and philosophically bears his loss! It is a stirring sight to watch these Eagles fly across the valley, or soar round and round in widening circles high above the mountain-tops; they imbue these desolate hills and heaths with life, and their barking cries are fitting music for such grand and rugged scenery.

Lower down the glen, on the face of one of those stupendous "storr" rocks which occur so frequently in the Highlands, a pair of Ravens have their nest. Although banished almost entirely from their lowland haunts, Ravens here are common enough; and this pair of birds have bred here for more years than the very oldest inhabitant of the scattered cottages around it can recall. The bulky nest, time-worn and bleached by many a tempest, is in full harmony with the sable owners. Season after season, often before the snow has melted on the hills, the old birds patch up and renovate their cradle, and with a regular "spring-cleaning," prepare it for the coming brood. The Raven is another of the very earliest birds to breed, and the eggs are laid on an average by the middle of March. Even a month previous to this date the Ravens are to be seen almost daily in and about their nest, repairing and cleaning it from time to time. This habit of visiting the nests long before the breeding season commences, and from time to time after it is all over, even during the late autumn and in midwinter, is a very curious one, and is common to nearly all the British members of the Crow tribe; Magpies, Jackdaws, Choughs, Crows, and Rooks all do so. This Raven's nest is a large and substantial structure, made of sticks, heather branches, and pieces of turf, thickly festooned with masses of sheep's wool, and lined with finer twigs, moss, dry grass, and wool, all felted firmly and smoothly together. The eggs are very small for the size of the bird, and various shades of green in ground colour, spotted and blotched with olive-brown and gray. The old Ravens are anxious enough at our intrusion, and croak their displeasure from the rocks, disturbing a pair of Peregrines, which a little later in the season will be bringing up their brood on the face of the same majestic cliff. A month later (in the middle of April) this fine bird lays its eggs. The Peregrine does not make much nest—a little hollow in a crevice of the cliffs, without lining or any other provision, is where the eggs are laid. These are four or five in number, and in colour exactly resemble those of the Kestrel, but are, of course, twice the size. The Peregrine is a noisy bird at the nest, and flies to and fro and round and round before the face of the cliff, chattering anxiously all the time. This beautiful Falcon feeds largely on Grouse and Rock Doves; and when its haunts are near the great bird bazaars, sea-fowl are taken in large numbers. I know of few finer sights in the bird world than the wild, impetuous swoop of the Peregrine upon its prey. Like a bolt from the blue sky, the Falcon hurls itself upon its unsuspecting victim, often tearing it in twain by the force of the swoop.

Most raptorial birds engage in various and beautiful evolutions in the air during the spring, and the Peregrine is no exception. The courtship of these birds is almost entirely carried on in the air. Kestrels will soar upwards until they look like mere specks against the background of glorious blue; Merlins and Hobbys will wheel and manœuvre in ever-widening circles above their breeding haunts; Eagles toy and buffet with each other high up in the sky; and the Peregrine and Buzzard oft mount in spiral course—evidently for pleasure—enjoying their honeymoon far up amongst the clouds. Aërial combats, too, are of frequent occurrence among these birds in early spring; and then their rapid movements and evolutions are truly astonishing. I have seen two male Eagles fight for victory in the tournament of love, now locked in tight embrace, anon flying upwards and upwards, each trying to get above the other, and then as rapidly descending, each striking its opponent with wings and claws, and finally disappearing behind the summit of a distant hill, probably to continue their fight in the aërial arena until one of the combatants was conquered.

On our way down the hills we come upon a curious assemblage of the Crow tribe. Five Hooded Crows, two Carrion Crows, and a Raven, startled at our approach, rise hurriedly from the broken ground below a steep cliff on the hillside. Where the Crows are, there will the carrion be. A dead sheep, in an advanced state of decomposition, is lying among the heather, and the foraging sable rascals, attracted by the stench, have swooped down upon the prize. I never see an assemblage of this kind without thinking of Waterton's masterly defence of the Vulture's nose. The Crows are equally gifted with great powers of scent, and can smell the carrion from afar. I have repeatedly known a dead sheep to be among the rocks until decomposed before the carrion birds discover it ; and I have more than once hidden such an animal among the long heather, and have never known these birds to visit or discover it until the effluvia has betrayed its whereabouts. My experience with the Griffon Vultures in Northern Africa also led me to the same conclusions. On the other hand, Crows are the most prying of birds, and often discover a carcase by accident whilst foraging among the hills. The Raven is the shyest of the party, and hurries off down the glen, grumbling to himself as he goes ; the Carrion Crows fly round and round, awaiting our departure ; but the Hoodies — impudent thieves and rascals!—almost allow us to get within gunshot before they fly reluctantly away, to settle a little distance off to watch our movements. These Hooded Crows are nesting in the glen. The big, bulky home of one pair is made in a stunted thorn-bush on the hillside, where every passer-by can see it, and where we can almost reach the contents without climbing at all. The female has only just commenced laying: a solitary egg, in form and colour exactly like those of the Rook, is all that the nest contains. This nest is an exceptionally early one, for the Hooded Crow is, for a mountain bird, rather a late breeder, and its laying-time does not begin, as a rule, until the first or second week in April. The Carrion Crow is a week or more later still. Both these birds, like the Raven, return to breed in the old nest, season after season, even in the face of much persecution and disturbance. The nest of each species very closely resembles that of the Rook.

From the cold, gray waters of the loch among the hills, the Diver's wild, unearthly scream sounds startling and clear. The fisher Osprey has not yet arrived from his winter haunts in South Europe and the basin of the Mediterranean; but pairs of Wild Ducks are paddling about the shallows, and their love-notes are borne by the breeze across the water. Where the hills fall sheer down in precipices to the sea-lochs, the White-tailed Eagle is breeding, making its eyrie in the steepest parts of the rocks. Almost any hour the big birds may be seen beating along the coast, or foraging inland in quest of any garbage or weakly, wounded bird or animal. This Eagle is very Vulture-like in its habits, and is also very fond of fish, preferring to prey upon the dead, stranded ones lying on the beach rather than to catch them for itself.

It is only the very earliest dawn of spring-tide yet among these upland solitudes; but most of the resident birds are either breeding or just about to do so. Sterile and desolate look the hillsides; cold and cutting the March wind; yet in another month a vast change will have come over the scene, and thousands of birds from the lowlands, and from across the sea, will have arrived for the summer, and the moors and mountains will echo their glad call-notes and joyful songs. Content with what we have already seen, we will defer a second visit until a more congenial season.

CHAPTER VI.

OUR FEATHERED ENGINEERS.

It is a well-known and profoundly interesting fact that many of the crafts practised by civilised men, as well as by savages, have been anticipated by animals much lower in the scale of organisation. We have many clever engineers and architects among insects, and even among fishes. Ants, the trap-door spider, the caddis, and the stickleback are amongst the most familiar examples; but in the present chapter we propose to confine our attention exclusively to the engineering capabilities of birds. Every one is familiar with the nests of birds. Even the most casual observer cannot have failed to notice the big nests of the Rooks high up in the lofty elms, the homes of the House Sparrows on buildings and in trees, or the curious cradle of the Martin attached to the walls of houses under the eaves. All nests, however, are by no means made on the same plan; the various raw materials are worked up in many different ways. For convenience of treatment we will divide the various examples of bird architecture into classes which will include all the most characteristic methods of building.

1. *Plasterers.*—In our first division, we will briefly glance at a few of the birds which are in the habit of mixing a portion of the nest materials into a rude kind of paste or plaster. One of the best examples of this peculiar class of architecture is the nest of the common Song Thrush, a bird as well known as it is appreciated for its sprightly form and charming song. The Song Thrush's nest undergoes two very distinct stages in the course of its construction. In the first place, the outside nest is formed of dry grass, scraps of moss, and sometimes a few slender twigs. This rather loose and flimsy nest is then carefully and compactly plastered with a thick coat of wet mud, worked well into the grass, and then the whole structure is finally lined with a thinner coating of rotten wood. This latter material is obtained from logs of wood or decayed stumps, those saturated with moisture being preferred. If none wet enough can be found, the birds moisten it in the nearest water, and with feet and bill work it on to the lining of mud, using their breast to finally smooth and round their beautiful handiwork. The whole structure is then generally left a day or so to partly dry, ere the first egg is deposited. The heat from the sitting bird soon completes the plastering process. We cannot help admiring the wonderful instinct which prompts the bird to select such a material for the final lining of its

nest. Wood, in preventing the escape of heat, is therefore the best material possible for the incubation of eggs, especially at a time when the atmosphere is chilly, and often charged with moisture; for the Song Thrush is one of the earliest birds to breed in the spring. All the other species of British Thrush plaster their nests, but not so much as the preceding bird. They make their nest in a precisely similar manner: first the grass stage, then a plastered coating of mud, but finish off with a final lining of dry grass. The Misselthrush perhaps uses more mud in the fabrication of its nest than any other member of this family. Another plasterer is the Kittiwake Gull. This bird builds its nest upon the jutting prominences of the steep ocean cliffs, forming it of turf and roots with the soil adhering to them, which are tramped and beaten with its wet feet, aided by the salt spray, into a mortar-like mass which in time becomes very solid and firm. Upon this foundation a further nest is made of seaweed, marine herbage, dry grass, and sometimes a few feathers. The Nuthatch is also very expert at plastering. This bird generally breeds in a hole of a tree, and usually plasters up most of the openings with clay, only leaving sufficient room for ingress. This curious little bird has been known to collect as much as eleven pounds of clay to plaster on its nest.

2. *Masons.*—From the plasterers we will now turn to the masons. Most interesting of these

are the House Martins and the Swallows. The former birds, as is well known, build their nests under the eaves of houses, in window-frames, beneath the copings of viaducts, in doorways, and even on the bare cliffs, both inland and on the coast. As in the Thrushes, the nests undergo several stages. The masonry part of the structure is first completed. The Martins, as soon as nest-building commences, may be seen alighting on the margins of ponds and streams, or even on the roads, to pick up little bits of mud. These pellets, one by one, are attached to the selected site on the wall or cliff until sufficient foundation is laid, when the real work of constructing the outer shell of the nest begins. Backwards and forwards the little masons fly, to and from the pools or roads, bringing each time a fresh piece of mud, which is carefully stuck upon the nest like so many bricks. The Martins only do a little at a time, leaving the work to set and become firmly cemented together, and to the wall, ere going on again. Gradually the shell is formed, neatly attached at the back, and sometimes overhead, to the supporting masonry, course succeeding course, a small aperture being left at the top for ingress. This semi-globular nest of mud is then lined with dry grass and feathers, and all is ready for the tiny white eggs. During dry seasons the Martins often have considerable difficulty in getting mud plastic enough for their purpose, and will fly long distances to obtain this material. Then, again, the mud sometimes will not adhere, and, time after time, the half-built structure falls, the patient architects beginning to repair the damage at once. House Martins are most interesting birds; and there are few prettier sights in spring-time than to watch them building their nests or repairing the old ones. Both birds assist in making the nest, and very often one will stay at home to guard and watch whilst the other seeks material. Swallows build in much the same manner, only they prefer a covered site for the nest, in a barn or other outbuilding. Sometimes it is built upon one of the beams supporting the roof, at others on a stone jutting out of a wall or chimney. The little bits of mud collected by the birds are formed into a shallow cup, and then lined with grass and feathers. The House Martin returns each year and uses its old nest; but the Swallow, though coming back again to its accustomed haunts, generally makes a new home close to the one of the previous season. Both birds assist in making the nest.

3. *Miners.*—The Sand Martin is certainly the best known little engineer in this department of bird architecture. The earthworks of the Sand Martin may be seen in almost every sandy cliff or steep railway cutting, more especially if such are near to water. The gravel pits and sand quarries are favourite resorts, and as the birds live in colonies, it is easy to watch their lively movements during the breeding season. With mar-

vellous perception the Sand Martins select the portion of the cliff where the ground is neither too hard nor too soft for boring. Regularly every spring-time the little engineers return to the old familiar colony; fresh nests are made in many cases by birds setting up housekeeping for the first time, by others whose homes have fallen in or been destroyed during the previous winter, or by those who change their quarters for no apparent reason to us. The tunnels formed by the Sand Martins are about three inches in diameter, sometimes more and sometimes less, the little architects scratching away the sand with their claws, their bodies working round like an animated drill, for two or three feet into the solid bank. Sometimes these passages turn and twist considerably. I have known them turn at right angles, and now and then two points of ingress are made to the same nest, whilst less frequently one entrance gallery will branch out into two passages, each leading to a different home. The Sand Martins, like the thorough engineers that they are, also provide for drainage by making the tunnel gradually slope upwards. At the end in a kind of chamber, the slight, slovenly nest of dead grass and feathers is placed. Sometimes the tunnels are deserted at a depth of a few inches or even more, the soil apparently being unsuited for boring; and very frequently a boulder or a large pebble stops the way, and the workings are deserted. Both birds work at boring these galleries, and the fine sand and soil falls in a heap on the ground below the entrance to them. Another bird miner, almost as well known as the Sand Martin, is the Kingfisher. This bird digs and delves into the steep banks of brooks and ponds, only the beak is the tool employed in boring instead of the feet. I have seen Kingfishers' beaks much worn after the nest has been made, owing to the hardness of the bank and the incessant way in which the birds have worked. The tunnel is made to a depth of three or four feet, and does not differ in any important particular from that of the Sand Martin. At the end of the burrow the eggs are laid on an accumulation of fish bones, for this bird, be it known, frequents its breeding hole on and off throughout the year. The gaudy Bee Eater, and occasionally the Roller, are also miner birds, but as they do not breed in this country, a detailed description is scarcely necessary. Another bird belonging to this class of architects is the comical Puffin or Sea Parrot. These birds love to breed on uninhabited islets where the soil is deep and easily burrowed, and in such chosen haunts the ground is literally undermined in all directions by their winding galleries. These tunnels resemble rabbit holes, and often extend for several yards through the soil ere the nest chamber is reached, which is usually lined with a little dry grass, and sometimes a feather or two, where the solitary egg is deposited. Puffins work with bill and claws, the latter being exceedingly sharp. In working, the birds throw the loose soil out behind them in an almost continual stream. Other miners are also found among the sea-birds, especially the Shearwaters.

4. *Wood Cutters.*—In the British Islands the birds falling under this category are few; but in the tropics there are a great many species, especially in the Picarian order of birds, which cut into wood for the purpose of making a nest. With us, the Woodpeckers are the most typical species, and are birds singularly well adapted by nature for cutting into timber, being armed with long, chisel-shaped beaks exceptionally strong and powerful. All the British species carry on their operations in a similar manner. Take, for instance, the Green Woodpecker. In spring this handsome bird selects a branch or portion of the trunk—usually one which is more or less decayed—and bores at first horizontally for a few inches, then the shaft is sunk in a perpendicular direction for a foot or more, the bottom of which being enlarged into a chamber, where the eggs are laid on the powdered wood without any other nest. Some of these Woodpeckers' holes are marvellous pieces of handiwork, beautifully smooth and round, and appearing as though they had been carefully cut out with a small gouge. The bill is used almost like a pickaxe, and every bit of detached wood is carried and dropped outside, where the accumulation below the tree often betrays the

presence of a tenanted nest. Our other feathered woodcutters are the Titmice. In every case these birds cut into decayed wood when they cannot find a hole ready made suited for their purpose. I have known the Coal Titmouse cut its way into a rotten stump in a hedgerow for a distance of eighteen inches, and make its warm nest at the bottom, of moss, wool, dry grass, feathers, and hair. The Blue Titmouse, though it often chooses a site in an old wall or a pump, frequently cuts its way into wood; and the Marsh Titmouse repeatedly does so. In boring, these little birds are very careful not to betray the nest by the fragments of wood cut out, and carry them away to a safe distance bit by bit. All the Titmice make warm and bulky nests in the holes they excavate or select. The Redstart and the Pied Flycatcher also breed in holes in timber, but never make them for themselves, always selecting one ready to hand.

5. *Felt-Makers.*—We now arrive at a class of birds whose architecture is of a far more elaborate kind than any hitherto treated with. Some of the nests made by the felt-makers are exceptionally elaborate and beautiful, the materials of the nest being felted or matted together with wonderful skill, and the structures themselves being remarkably handsome in design. Our best-known felt-maker is the Chaffinch. As soon as spring-time bursts the hawthorn buds, the Chaffinches may be found at work on their beautiful abode. A site

for the nest is chosen in some suitable crotch of the branches, often one covered with tree-moss and lichen being selected. A dozen nests of the Chaffinch lie before me as I write, selected with care for the variety of their materials and the handsome manner in which they are made. In most of them moss and fine grasses are used; in a few, moss alone; these materials being worked and felted together with spiders' webs, bits of vegetable down, and lichens. The outside of the nest is made to closely resemble in colour the surrounding objects, whether branch, or trunk, or foliage; the inside is warmly lined with hair, feathers, and down from various seeds. Chaffinches are most fastidious birds during the period of nest-building; very anxious, too, for their uncompleted home, and do comparatively a small piece of work each day. A well-made, handsome nest takes a fortnight to build; but nests made by young birds are more careless in execution, and sometimes put together in a week. The female is the builder, the male bringing most of the material. Some nests are very beautifully garnished with gold or silver lichen, bits of paper, flakes of elm bark, and even the prismatic wings of insects mixed up with spiders' webs. Another skilful felt-maker is the Dipper. Greenest moss is almost exclusively used in forming the outer portion of the nest, this material being cleverly felted together into a compact globular mass, and lined with dead leaves, dry grass, and moss. It is usually built in a crevice close to the stream—sometimes actually behind a cascade, in a nook of the tree-roots, where the humidity of the situation keeps the external material fresh and green. The Great Titmouse is another clever worker in felt, sometimes making a ball-like nest of moss, and hair, and wool, lined with feathers, in an old squirrel's drey, or in the deserted nest of a Crow or Magpie. The Long-tailed Titmouse, one of the smallest of our British birds, also makes a felt-like nest, spherical in shape, with a hole for ingress on one side near the top. I have a nest of this bird with two holes, one of them provided with a flap or trap-door of felted moss, which opened and closed as the birds went in and out. The materials used are precisely similar to those selected by the Chaffinch; but the nest, instead of being on the larger branches or in a crotch, is usually placed among the more slender twigs often of the holly or the prickly gorse. The Wren also ranks as a felt-maker, some of its nests being so strongly put together as to require considerable force to pull them in pieces.

6. *Weavers.*—The birds in this division are remarkable for the skill and dexterity with which they fabricate their nests out of various textile materials, rarely if ever using twigs or sticks in their construction. Best known of all the weavers is the House Sparrow. Singularly enough this bird makes two very distinct types of nest—one a slovenly structure in holes of trees and buildings,

and the other domed and intricately woven in the branches of trees and bushes. The Sparrow's powers of weaving are admirably illustrated in this latter form of nest. Grass, straws, and very often twine and worsted, are all woven together so skilfully that it is a difficult matter to pull the nest to pieces. This outer structure is further lined with quantities of feathers, wool, and any soft material the birds may chance to find. I have even seen them pulling hemp from the frayed end of a clothes-line. Both birds build the nest and collect materials. Another clever weaver is the Golden Oriole, which suspends its beautiful nest between a horizontal fork of a branch, the various materials—sedges, dry grass, and leaves, and often scraps of newspaper—all being deftly woven. The materials of the rim of the nest are wound round and round the supporting branches, strips of bass often being used for the purpose. There are many other birds to be classed among the present division; but as they are most of them inhabitants of foreign lands a detailed description of their handiwork is scarcely required. Specially interesting, however, are the Hang-Nests of the New World, and the Weaver Birds, and many small Finches of Africa and India. Some of these pensile nests are remarkably beautiful and curious, being suspended from slender branches, often over water, where all save winged enemies are set at defiance. The Buntings, Larks, Pipits, and Chats, are all clever weavers.

7. *Basket-Makers.*—By far the greatest number of nest-building birds in the British Islands come under the present category. Among the basket-makers must be included the Warblers, the Crows, the birds of prey, the Pigeons, the Herons, and the Cormorants. All these birds follow the principle of basket-making in constructing their nest, winding the sticks, twigs, and grass-stems in and out, across and across, very similar to the withes of a basket. Take, for instance, the delicate flimsy nest of the Whitethroat. It is made principally of round, dry grass-stalks, each wound in and out with great skill and regularity, the whole structure being finished off with a lining of horsehair. The Blackcap and the Garden Warbler are other instances. The Crows, several of the birds of prey, as for instance the Sparrowhawk, and the Herons, all make large basket-like nests of sticks, so strongly put together that you may stand upon them in perfect safety. One of the best instances of this peculiar class of bird architecture is the wonderful abode of the Magpie. Every one knows the bulky nest of this pretty bird, roofed over with a canopy of basket-work which almost defies destruction. I have a vivid remembrance of this, for when a boy it was an agreement between the keeper and myself that I had to pull out all the nest from the tree as a consideration for the privilege of being allowed to climb to the nest and take possession of the contents! Very often this was no joke, for I have seen Magpies' nests which

weighed upwards of half a hundredweight, massive and strong, the accumulation of years. Slighter made; yet equally interesting, are the wicker-work platforms of sticks made by the Pigeons. Though frail in appearance, they are remarkably strong and well put together, and the cake of excrements that accumulates under the nestlings increases the durability of these nests.

8. *Scaffold-Builders.*—The birds in this division really combine with the basket-makers in engineering skill, forming a nest very similar to the other Warblers, but placing it on reeds. The Reed Warbler is the best known of this group. It selects three or four convenient reeds, and using them like scaffold-poles entwines its nest, basket-like, round and round them. This nest is composed of dry grass, broad leaves of the reeds, and rootlets—the latter material also forming the lining.

9. *Raft-Makers.*—In our next division we will briefly glance at those birds which make a floating nest—literally a raft—on which to hatch their young. The best-known species in the present group is most probably the Moorhen. This bird frequents most ponds and reedy pools where the vegetation round them is sufficiently dense to afford the necessary cover. The Moorhen sometimes makes its nest on dry land, even on a flat branch of a tree; but it usually builds among the rushes, iris, and mares'-tails, some distance from the shore. Here a large heap of

rubbish is collected—rotten flags, grasses, and aquatic vegetation of all kinds—anchored safely to the reeds and rushes. On this heap of refuse a dryer stratum of grass, dead leaves, and bits of reed stems is formed into a shallow nest in which the eggs are laid. The Coot's nest is usually built in a similar situation, and nearly of the same materials—a large, floating raft of aquatic vegetation. The last of these curious nests that we will notice are those of the Grebes. Many of these structures are rafts of vegetable *débris*, often quite unattached to any rush or reed, and absolutely floating free upon the surface of the water. All these raft-builders are in the habit of adding fresh material to their nests from time to time, to repair the damage caused by the washing of the water.

10. *Upholsterers.*—The birds which come into the present group comprise the Ducks and Geese. These birds possess the singular habit of padding their nests with down from their own bodies as the eggs are being laid, so that, by the time incubation commences, the structure is warmly lined with a bed of gossamer lightness. Man has taken a lesson from the birds, and forms from the down of the Eider Duck the luxurious coverlets which are so highly prized for their warmth and exceeding lightness. Typical of the Ducks, we may briefly glance at the nest of the Wild Duck and of the Eider. The first-named bird by no means always builds its nest near the water; sometimes

it is a mile or more from the pool. It is usually a little hollow, lined with dry grass, moss, and leaves, and the down from the parent's body is added last of all. The Eider Duck always breeds near the sea, usually making a slight nest of marine herbage in a crevice of the rocks on an uninhabited island, and the warm padding of down is gradually added as egg after egg is laid. Some Ducks breed in holes in trees, as the Golden-eye; under rocks, as the Merganser; and in burrows, as the Sheldrake; but all upholster their nests in the same singular manner. The Geese make bulky nests of dead grass and leaves, heather, and aquatic vegetation, lining them with moss, adding the final upholstering as the eggs are being laid.

11. *Tailors.*—Although we have no representative of this class of feathered engineers in the British Islands, I make no apology for introducing them into the present chapter. One of the most remarkable and curious examples of bird architecture in the whole world is the nest of the Indian Tailor Bird (*Orthotomus longicauda*). This interesting little bird selects a broad leaf of some tropical plant, and draws the edges together into a cone, which is securely fastened with a thread of vegetable fibre. This cone is then lined with fine, dry grass and scraps of vegetable down. But the most curious part remains yet to be told: the thread of fibre which sews the leaf together is absolutely *knotted*, just as a tailor would do his thread! The Tailor Bird is a little, brown, insig-

nificant creature, yet the masterly way in which it forms its wonderful nest cannot fail to excite the highest admiration.

I should here also remark that many birds are masters of several of the crafts here enumerated. The Magpie, for instance, as well as being an expert basket-maker, is also a plasterer, and many pounds weight of mud are used by this bird to cement the foundation of sticks together, a lining of that material being used before the final one of fine roots. The Titmice, many of them, as well as being wood-cutters, are felt-makers too, their nests in the hollow stumps being beautifully felted. With this we will conclude the brief notice of our feathered engineers. Many curious and beautiful nests there are, especially in the tropics; yet most, if not all, are made on one or the other plan which we have already noticed. Birds' nests ever excite admiration, even in the most casual observer of Nature's wonders; and the variety in the method of their construction, in the selection of materials, in the sites they occupy, and, above all, the mental powers called into play by the birds themselves in fabricating them, make them a fascinating subject for contemplation, and furnish abundant material for patient, loving study, and research. Spring-time is a season of birds' nests, and the naturalist cannot be better occupied than in working out the secrets which surround these beautiful objects.

CALENDAR FOR SPRING.

Species.	March.	April.	May.
Golden Eagle . . .	Breeding	Breeding	Breeding
White-tailed Eagle .	,,	,,	,,
Peregrine Falcon. .	—	,,	,,
Hobby	—	Arrives	,,
Merlin	—	,,	,,
Kestrel	Many arrive	Pairing	,,
Honey Buzzard . .	—	—	Arrives
Marsh Harrier . .	—	—	Breeding
Hen Harrier . . .	—	Arrives	,,
Montagu's Harrier .	—	,,	,,
Sparrowhawk . . .	—	Breeding	,,
Common Buzzard .	—	,,	,,
Barn Owl	—	,,	,,
Long-eared Owl . .	—	,,	,,
Tawny Owl . . .	Breeding	,,	,,
Missel-thrush . . .	,,	,,	,,
Song Thrush . . .	,,	,,	,,
Blackbird	,,	,,	,,
Ring Ousel . . .	—	Arrives	,,
Dipper	In song	Breeding	,,
Wheatear	Arrives	,,	,,
Whinchat	—	Arrives	,,
Stonechat	In song	Breeding	,,
Redstart	—	Arrives	,,
Robin	Breeding	Breeding	,,
Nightingale . . .	—	Arrives	,,
Whitethroats . . .	—	Arrive	,,
Blackcap	—	Arrives	,,
Garden Warbler . .	—	—	Arrives
Reed Warbler. . .	—	—	,,
Sedge Warbler . .	—	Arrives	Breeding
Grasshopper Warbler	—	,,	,,
Chiffchaff	—	,,	,,
Willow Wren . . .	—	,,	,,

Species.	March.	April.	May.
Wood Wren	—	Arrives	Breeding
Goldcrest	In song	Breeding	,,
Great Titmouse	,,	,,	,,
Blue Titmouse	,,	,,	,,
Coal Titmouse	,,	,,	,,
Marsh Titmouse	,,	,,	,,
Hedge Accentor	Breeding	,,	,,
Wren	In pairs	,,	,,
St. Kilda Wren	—	—	,,
Creeper	In pairs	Breeding	,,
Nuthatch	—	,,	,,
Raven	Breeding	,,	,,
Carrion Crow	—	,,	,,
Hooded Crow	—	,,	,,
Rook	Breeding	,,	,,
Jackdaw	—	,,	,,
Magpie	In pairs	,,	,,
Jay	,,	,,	,,
Red-backed Shrike	—	—	Arrives
Starling	In pairs	Breeding	Breeding
Goldfinch	In song	In pairs	,,
Siskin	—	Breeding	,,
Greenfinch	—	,,	,,
Hawfinch	—	In pairs	,,
House Sparrow	In pairs	Breeding	,,
Tree Sparrow	—	,,	,,
Chaffinch	In pairs	,,	,,
Linnet	,,	,,	,,
Lesser Redpole	—	—	In pairs
Twite	In song	In pairs	Breeding
Bullfinch	—	Breeding	,,
Corn Bunting	In song	In pairs	,,
Yellow Bunting	In pairs	Breeding	,,
Cirl Bunting	—	In pairs	,,
Reed Bunting	In pairs	Breeding	,,
Pied Wagtail	Many arrive	,,	,,

CALENDAR FOR SPRING.

Species.	March.	April.	May.
Gray Wagtail . . .	At breeding grounds	Breeding	Breeding
Yellow Wagtail . .	Arrives	,,	,,
Meadow Pipit. . .	—	,,	,,
Tree Pipit	—	Arrives	,,
Rock Pipit. . . .	In pairs	Breeding	,,
Skylark	,,	,,	,,
Wood Lark . . .	Breeding	,,	,,
Swallow.	—	Arrives	,,
Martin	—	,,	,,
Sand Martin . . .	—	,,	,,
Swift	—	,,	Arrives
Goatsucker. . . .	—	—	,,
Kingfisher	In pairs (young)	Breeding	Breeding
Green Woodpecker .	,,	,,	,,
Great Spotted Woodpecker	,,	In pairs	,,
Lesser Spotted Woodpecker	—	,,	,,
Cuckoo	—	Arrives	,,
Ring Dove. . . .	Begin cooing	Breeding	,,
Stock Dove . . .	In pairs	,,	,,
Rock Dove . . .	Few breeding	,,	,,
Turtle Dove . . .	—	Arrives	In pairs
Ptarmigan	In pairs	In pairs	Breeding
Red Grouse . . .	Breeding	Breeding	,,
Black Grouse . . .	—	Pairing	,,
Capercaillie . . .	—	Breeding	,,
Pheasant	Begin pairing	,,	,,
Partridge	,,	,,	,,
Quail	—	—	Arrives
Heron	Begins breeding in south	Breeding	Breeding
Corn Crake . . .	—	Arrives	,,
Spotted Crake. . .	—	In pairs	,,
Water Rail. . . .	In pairs	Breeding	,,

Species.	March.	April.	May.
Moorhen	In pairs	Breeding	Breeding
Coot	Flocks disperse	In pairs	,,
Stone Curlew	—	Arrives	,,
Oystercatcher	—	In pairs	,,
Ringed Plover	In flocks	Flocks disperse	,,
Golden Plover	,,	,,	,,
Lapwing	Flocks disperse	Breeding	,,
Red-necked Phalarope	—	—	Reach breeding grounds
Curlew	Leave coast	In pairs	Breeding
Whimbrel	—	—	Arrives on coast
Common Sandpiper	—	Arrives	Breeding
Redshank	Trills at breeding places	Breeding	,,
Greenshank	—	Arrives	,,
Dunlin	—	Disperse & leave coast	,,
Sanderling	Passes on migration	Passes on migration	—
Woodcock	,,	Breeding	Breeding
Snipe	—	,,	,,
Sandwich Tern	—	Arrives	,,
Common Tern	—	,,	,,
Arctic Tern	—	,,	,,
Lesser Tern	—	—	Arrives
Black-headed Gull	Return to breeding places	Breeding	Breeding
Common Gull	—	Returns to breeding places	,,
Lesser Black-backed Gull	—	,,	,,

CALENDAR FOR SPRING.

Species.	March.	April.	May.
Great Black-backed Gull	—	Returns to breeding places	Breeding
Herring Gull	—	,,	,,
Kittiwake	—	,,	,,
Great Skua	—	,,	,,
Richardson's Skua	—	—	Returns to breeding places
Puffin	—	—	,,
Razorbill	—	Returns to breeding place	Breeding
Black Guillemot	—	—	
Common Guillemot	—	—	,,
Red-throated Diver	—	Returns to breeding place	,,
Black-throated Diver	—	,,	,,
Great Crested Grebe	—	At breeding grounds	,,
Little Grebe	Breeding	Breeding	,,
Manx Shearwater	—	—	,,
Fulmar	—	—	,,
Stormy Petrel	—	—	,,
Fork-tailed Petrel	—	—	,,
Gray-lag Goose	Returns to breeding places	Breeding	,,
Sheldrake	—	,,	,,
Wigeon	Return to breeding places	In pairs	,,
Teal	—	—	,,
Garganey	—	—	,,
Shoveller	—	—	,,
Mallard	Return to breeding places	Breeding	,,

Species.	March.	April.	May.
Pochard	—	—	At breeding grounds
Tufted Duck . . .	In pairs	Breeding	Breeding
Eider Duck . . .	Flocks disperse	In pairs	,,
Red-breasted Merganser	,,	,,	,,
Gannet	Return to breeding places	Building	,,
Cormorant . . .	—	,,	,,
Shag	—	,,	,,

SUMMER.

Part II.—Summer.

CHAPTER I.

THE WONDERS OF THE SUMMER.

> From brightening fields of ether fair disclos'd,
> Child of the Sun, refulgent Summer comes,
> In pride of youth, and felt through Nature's depth:
> He comes attended by the sultry hours,
> And ever fanning breezes on his way;
> While, from his ardent look, the turning Spring
> Averts his blushful face, and earth and skies
> All smiling to his hot dominion leaves.

W<small>HO</small> shall say when the glories of spring merge themselves into the wonders of summer? So gradually does the change occur, that each is beautifully blended in the other; Nature abhors hard and fast lines, and even seasons as well as many species merge insensibly together. However, in the early days of June we may safely conclude that summer has commenced. Life now in its myriad forms is rapidly approaching the very zenith of its splendour; the pageant of the year will shortly be in fullest gala prospect. No matter where we may wander at this delightful season, the same luxurious wealth and variety of life is presented in the plant as well as in the animal world. This glorious pageant of Life is endless; from morning until evening, and from evening until dawn, the wild denizens of the woods and fields, the mountains and the waters, are all astir —each creature has its own particular time of activity. No matter which hour of the twenty-four we may stroll abroad, the same restless activity prevails—in bird life, perhaps, most apparently of all.

Go out among the birds at very earliest dawn of day. Not even the faintest streak of light yet glimmers in the eastern sky, still the Thrushes are awake; and you may hear the speckled songsters begin their morning anthem, even though the birds themselves are hidden in the gloom. The Robin sings long and loudly from the bushes, and the restless Wren chants cheerily from the briars and thickets. Then watch the rosy streaks of morning spread athwart the sky; see the sun at last appear, the narrowest of crescents, above the green expanse of forest, and slowly rise in glorious splendour, dispelling the night mists from the hillsides, and waking all diurnal creatures from their short and fitful sleep. How cool and fresh all Nature seems in the first few hours of these summer mornings! What a matchless concert of wild melody swells through the woods and groves; what flitting of busy wings amongst the foliage; what hurrying to and fro;

what animation and excitement! Singing and love-making, bathing and feeding, fill the early hours of day. Everywhere among the trees, and in the hedgerows, and amongst the herbage, crowded nests of young birds are clamouring for sustenance, and brooding birds are being fed by their faithful mates. The woods, and groves, and the wide expanse of deep blue sky resound with song, each bird apparently striving to out-sing all others within hearing. One by one the tuneful Skylarks flutter from the dripping clover, and rise in spiral course, bursting with song. Birds are everywhere; they sing the time of Summer's triumph, and fill each hour of morning with their revelry and joy.

But as the day advances and the brilliant sun rises higher and higher in the heavens the heat increases, and gradually a lull and languor creeps over most living things. Songster after songster ceases its lay and finds a refuge among the cool green leaves and in the shady nooks; the Larks drop down again into the tall and juicy aftermath, and quietness almost reigns supreme. This stillness is even more intense as the noonday heat becomes more and more oppressive. Now is the time that insect life is most astir. From the fragrant lime trees spreads a dreamy hum of sound as the bees garner the plenteous feast; gay painted butterflies dance among the flowers in the open; they love the sunshine, disappearing when the clouds obscure his rays, flitting out again when they have passed; whilst flies and gnats and other winged creatures hurry to and fro in countless hosts with noisy buzz, and drone, and hum. Little else but the Swallows and Martins in their glittering steel-blue livery venture out into the noonday heat; but these birds feed on insects and follow their prey abroad, chasing them round the limes, under the oaks and beeches, circling above the poplars, and coursing lightning-like beneath the drooping branches and out into the open meadows. The hush that settles over the haunts of animated Nature during these blazing hours of a summer day is most impressive. In the cool shade of the surrounding greenery most of the feathered tribes are skulking; too languid to sing, many of them too lazy to feed. Here and there a bird may be seen flitting among the leaves, its wings glinting fitfully in the stray sunbeams, which play like delicate embroidery on the foliage. Now and then a chirp or startled cry relieves the noonday silence. The gurgling and the splashing of the trout-stream, as it falls over the moss-grown weir, sounds soothing to the ear; and its limpid waters are as refreshing to the eye, as they flow silently along through the meadows like a streak of silver, and under the alder and ash-trees by the coppice, where the cattle, knee-deep in the beck, stand half-asleep, impatiently lashing with their long tails the flies that torment them. The greenness of the country-side is almost universal. The sun has not yet scorched the pasture-fields, and every bush and tree is clothed in its summer robe of darkest green. The spring-time blossoms have faded; scattered to the winds are the fleecy branches of hawthorn; the pink-and-white glory of the crab, and the snow-white clusters of the bird-cherry were ruthlessly strewn under foot long ago. Nature produces lavishly, and just as lavishly destroys. No living things escape the doom of decay which is the penalty of their being. They bud, and bloom, and shine in pristine splendour for a season; then they wither and fall, and are gone. But the hedges just now are spangled with wild roses, and the fragrant honeysuckle twists, and twines, and hangs in tempting clusters from the taller bushes, loading the air around with perfume of rarest sweetness. Few buttercups, like cloth of gold, glisten in the sunlight now, and the daisies have vanished or are hidden in the long grass; but the tall "moonpennies," or "ox-eyes," gleam in the meadows high above the herbage; and the fragrant meadowsweet and graceful foxgloves are scattered along the hedgerows. Summer blooms replace the sweet, fair flowers of spring; the blue and white carpets of hyacinths and anemones have vanished; but the pink clover and the vetches set the fields aflame with colour and saturate the air with scent. In amongst the clover the leverets sport and play, and the rabbits towards evening regain their accustomed activity, and venture from their burrows in the steep banks between the coverts and the open pasture-grounds.

The sun in his blazing path across the heavens, is already descending through the western sky. The heat and oppressiveness of the summer day are past. Gradually the various wild creatures come forth, and the bustle and activity of morning are renewed. Long shadows creep across the fields—longer and longer as the sun approaches the horizon. Once more the Thrushes lead off the feathered orchestra, and the crash of melody, sweeter far than any organ music, peals in fitful clashes, and fills the woods and groves with song. Life in its endless forms musters once more into review order, and flower, and insect, bird, and animal, take up their places in the stirring pageant. Now the slanting sunbeams play upon the gray trunks of the ash trees, and light up the silver bark of the beeches, and glint upon the birch stems. Now and then, athwart the rays of light in the open glades, the Brown Flycatcher flits in chase of insects, and the Chiffchaff shouts incessantly from amongst the shimmering leaves. It would be vain to attempt to catalogue all the living wonders abroad during a summer evening, or to enter into details of their various ways, in the space of a chapter or even of a volume. Right through the fading hours of daylight, the wild inhabitants of the fields and forests are all astir; but as the sun drops down behind the western woods, bird after bird hurries off to its retreat, and the evening plethora of song grows more and more fitful as night steals back again. Then is the season for another class of creatures to bestir themselves. The various animals—the mice and rats, the hedgehogs, stoats, weasels, otters, hares, and rabbits, are all fond of the night, and delay their gambols until dusk. You may hear them on every side, frolicking with each other, marauding after prey, bold yet timid, venturesome yet shy.

Then, with approaching darkness, the Nightjar steals forth, and the Landrail grates his music from the meadows. All night long the plaintive Nightingales hold their concert, and Reed Warblers and Sedge Birds warble fitfully from the vegetation near the stream. They seem too restless to sleep. The comparative stillness of the woods, the fragrance cast off by sleeping trees and flowers; the scent of hay, and lime, and meadow-sweet, the freshness and coolness after the noonday heat, make nights of midsummer deliciously sweet and soothing. You may sit and muse in the woods for hours together on such nights as these—sit and ponder over the mysteries of the life, waking and sleeping, everywhere around you; and when gazing up into the starry sky, across which the summer meteors flash at intervals, your thoughts may well embrace the higher questions still—the presence of other beings far out yonder in the spangled firmament; the Universality of Life away from our own small yet glorious planet, which circles through that space, on which we live and move, and have our birth, and life, and death. For it is no more rational to conclude that Earth is the centre and sole supporter of Life, than it is to presume that that Life which animates even the tiniest atom of organic matter is not as immortal and eternal as the grand and infinite Universe of which it forms so insignificant a part.

Night is favourable to such thoughts as these—the darkness and the stillness bring man closer to the vast unseen and inconceivable Power which behind this mighty Universe controls its working and shapes its destiny. A little Sedge Bird, in the spinney yonder by the stream, wakes the stillness of this beautiful night; it warbles a moment or two, then all is silent once more. Now what is the history of that feathered mite? Its song awakes a train of thought that leads us back into the misty past—back to the beginning of its existence as a bird—back to earlier ancestors still—back to the glorious dawn of Life upon this fair world of ours. Think of this Sedge Bird's long, eventful history as our planet has gone cycling on through the vast uncounted ages which bring this little Warbler from those dark, mysterious times down to to-night. What is this power of progressive development which seems ever to tend from a lower to a higher stage of existence which man in his wisdom has termed Evolution? Who the designer, and What the potent force which controls this beautiful process? Hark! there is our Sedge Bird warbling again in the spinney—he is one fraction of a group which comprises the highest developed of all known birds. Focussed in the Oolitic ages we see the fossil Archeopteryx the branch of life from which the twig of bird life seems to have sprung. What a link with the past even this little Sedge Bird makes! Then think of the countless other forms of life, each with a long and eventful history behind it, each the result of this grand plan of development from a lower to a higher, from a simple to a complex. How much more interesting does the study of Nature become when we attempt to peer into her secrets and her history from such a starting-point as this! Then the question arises: Has Life reached its appointed standard of perfection yet? Is development still in progress, and what is its ultimate destiny? It may be that this world of ours has nearly reached the limits of development, and that all things are about to fade away through future ages as insensibly and as beautifully as they appeared; for it is only reasonable to suppose that Planets, like all other objects of the Universe, are called into being, live and flourish, decay and die, such being the grand law which appears to govern all things.

But the brief night speeds rapidly away, and the dawn once more is being heralded by the various creatures of the day. The wonders of summer, and the grand questions they so beautifully illustrate, are endless to a meditative mind, and afford a rich field indeed for observation, contemplation, and research.

CHAPTER II.

AMONG THE BIRDS IN SUMMER.

BIRD LIFE endows a northern summer with much of its fairest charm. The quiet, dreamy beauty of our English woods and fields at this delightful season is imbued with that sense of joyous life the wild birds give. We miss much of the unusual activity which prevailed among the birds in spring; their crash of vernal melody is spent, and the music of the woods and fields, if more universal, is toned down and softened by the voices of little songsters from across the sea. The loud, powerful notes of the Thrushes are now varied with the voices of the Warblers, and the harsher cries are mellowed by the call of the Cuckoo and the murmur of the Turtle Dove. As we saw in spring, this latter bird is one of the latest of all our summer migrants. It loves the cover of the deepest woods, is shy and timid, yet garrulous enough during the early days of summer, when love and courtship are in progress. Another bird of summer is the Swift—one of the last to come in spring and one of the first to leave in autumn.

[margin: Turtle Dove laying, 5th June.]

Summer is the season chosen by most of the migratory birds for nesting duties, when insect life is abundant.

It may, perhaps, be as well to take a hurried peep at the domestic arrangements of some of our commoner birds, ere visiting the nests of those only found in remoter districts. One of the most beautiful nests to be met with amongst our southern fields in early summer is that of the Red-backed Shrike. It is a large, bulky structure, usually placed in a lofty bush or hedgerow, and is made of the dry stalks of plants, grass, roots, and moss, and lined with hair and wool. The eggs are five or six in number, and remarkable for their great diversity of colour. In the orchard trees, especially those trained along the wall, or in a chink of the bark of some rugged elm or oak, we may often find the nest of the Spotted Fly-catcher. It is a beautiful little structure, made of grass, and moss, and roots, cemented with spiders' webs, and lined with wool, hair, and feathers, in which the female lays her half-dozen greenish-blue eggs, thickly marbled with brown. In early summer the Redstart, too, is busy bringing up its brood in a nest in some hollow stump, or in a crevice of a wall, its delicate blue and spotless eggs being exceptionally beautiful objects. In the tangled hedges, and amongst the luxuriant growth of vegetation, in the woods, and by the trout-streams, our delicate summer Warblers are busy bringing up their broods—noisy Whitethroats,

Red-backed Shrike nesting, 3rd June.

Spotted Fly Catcher has eggs, 10th June.

Redstarts nesting, 1st June.

dulcet Blackcaps, shy Garden Warblers, and Sedge Birds; and, where the bushes and brambles are thickest, the skulking Grasshopper Warbler builds its nest, and chants its monotonous song at all hours of the day and night. This latter species is most interesting; no other British bird is more shy and retiring. You may hunt it up and down the cover, through thicket after thicket, along the hedge bottoms, and the tangled, matted grass and briars, without ever once getting a glimpse of it; all the time its sibilant music betraying its constantly changing whereabouts. Common Buntings and Willow Wrens, Greenfinches, and Swallows, Martins, and House Sparrows are all now deeply engaged in family duties. Now is the time to search for the charming little nest of the Lesser Redpole, snugly placed in a crotch of the hedgerows, or in the branches of the young larch trees in the plantations. It is a cosy, yet a tiny home indeed, made of moss and dry grass, bound together with a few roots and twigs, and warmly lined with feathers and down from the willow-tree and other plants. The eggs are equally beautiful—greenish-blue in ground colour, spotted with purplish-red, and sometimes streaked with darker brown.

All laying between 1st & 11th June.

Redpoles have eggs, 5th June.

During the early weeks of summer, vast numbers of young birds leave their nests to make their first appearance among the trees and bushes. Many of these are second broods. By the end of June the young Pied Wagtails are strong on the wing, and may be seen in the turnip-fields still attended by their parents. The first broods of young Greenfinches are out by the same time. On the grass-fields flocks of young Starlings congregate, in many cases attended by a few old birds; and families of Long-tailed Titmice may be met with in the woods and along the hedgerows at the same period. Small flocks of House Sparrows repair to the turnip-fields and the meadows, where they subsist chiefly on seeds.

Starlings flock, 13th June.

Sparrows flock, 23th June.

During the course of our summer rambles we meet with many little feathered strangers. Our old friends, the resident birds, we look to meet as a matter of course; but every now and then some rare and interesting summer visitor comes before our notice. As we wander through the fields by the sides of the hedges, the garrulous little Whitethroat takes his short, hurried flights before us—now down into the brambles, then up into the air, singing lustily as he goes; and every now and then the Cuckoo's gladsome notes sound full and loud from the woods. Very Hawk-like in appearance, this latter bird may now be seen flying from tree to tree, uttering a chuckling kind of cry. The hens are busy prying about in all likely places in search of nests in which to lay their alien eggs; the notes of the male are now not so rich and clear, and are often composed of three syllables—*cuck-cuck-oo*. Along the quiet reaches of the stream or round the margin of the lake, we may meet with the Common Sandpiper or

Cuckoo's notes change, 15th June.

Sandpiper sitting, 1st June.

Summer Snipe, a little wading bird that retires to the muddy coasts of South Africa during winter, visiting our northern waters to rear its young. It runs daintily along the shore, or even perches on a boulder in the stream, making its slight nest under a little bush near the water, in which it lays its four pear-shaped eggs, creamy buff, spotted with dark-brown and gray.

In the pastures, where the lazy cattle are grazing or standing in the cool shade of the spreading trees, impatiently lashing their tails or turning their heads from side to side, to rid themselves of troublesome flies, we are sure to find the dapper little Yellow Wagtail. The nest is made under the shelter of yonder hedgerow, and the parent birds come hither to catch the insects. See how daintily they run about among the grazing animals, close to their very mouths, busy in search of food. High in the blue sky the Swifts are darting up and down, screaming as they fly; whilst in the lower atmosphere the Swallows and House Martins are coursing to and fro, twittering to each other in their joy. Round and round the cattle they fly, sweeping under the branches, busily ridding the poor tormented animals of their insect plagues. The Tree Pipit now sings rapturously; and now and then you may chance to come upon a brood of young Partridges and their parents, especially in the quiet corners of the fields near the gateways where the anthills are often seen. It is a pretty sight to watch

Yellow Wagtails, young fledged, 10th June.

Swifts laying, 3rd June.

Partridges hatching, 20th June.

these active little creatures following their parents; but as the wandering Hawk crosses the sky their ever-watchful mother gathers her brood together with a warning cry and shelters them beneath her wings. Amongst the corn which is now just shooting into ear the Quails are nesting. These birds are polygamous, one male mating with several females, which often lay in the same nest. Usually several nests may be found close together in the same patch of corn or "seeds." In the woods the broods of young Pheasants are rapidly advancing to maturity, guarded by the ever-watchful keepers from Crows, and Hawks, and other predatory creatures. Whichever way we chance to turn, birds are sure to be seen; but it is interesting to notice how quickly they disappear at the approach of the heavy thundershowers. As soon as the first warning drops patter heavily on the broad leaves bird after bird seeks shelter amongst the densest foliage, and rarely one ventures forth until the heavy rain has ceased. Sometimes, however, the Swallows and Martins keep the air, and career about in the drenching rain without any inconvenience. Their dense, glossy plumage seems impervious to the water, and they flit about all indifferent to the storm. But as soon as the rain has ceased, and the sun shines brightly forth again, the birds hop out from their retreats, many of them bursting into song. All is gladness once more, and the parched, thirsty earth, and dusty, drooping vege-

Quails laying, 5th July.

tation, now refreshed and beautified, are full of rare fragrance.

On the quiet pools, whose margins are fringed with a dense bed of flags and rushes, the Waterhen finds a home congenial to its taste. This bird may often be seen walking about the short grass near the water, to which it instantly retires when alarmed. It makes its bulky nest among the rushes, and often rears as many as three broods in the season. The chicks are covered with down as black as jet, and are able to swim and dive almost immediately after they leave the egg. The Waterhen is a careful mother and leads her numerous family about the pool searching for food. When tired the chicks may often be seen resting on the broad, flat leaves of the "can-dock," and may sometimes be watched chasing an insect across them. Waterhens often perch in trees, and swim and dive with admirable grace and quickness, although their feet are not webbed, or even lobed, like those of the Coot. By the water-side another interesting little bird may often be met with. This is the Reed Warbler. It loves the reed-beds and osier-thickets, and is so skulking in its habits that it is rarely seen, only betraying its presence by its song. It is a most industrious songster, and is, therefore, not easily overlooked. As you wander along the fringe of tall reeds by the sluggish stream, you may see the slender stems quiver as the little reed-bird hurries through them with

Young Waterhens abroad, 3rd June.

great celerity, and now and then you are greeted with bursts of defiant song. By using the greatest caution in parting the reeds aside you may perchance be fortunate enough to see the little brown musician clinging to the swaying reed. The most casual search amongst the reeds will reward you with a sight of the nest of this Warbler. It is suspended on three or four reed stems several feet above the water, and is wafted about by every breeze that disturbs the reeds. The nest, as we learnt in spring, is made of coarse, dry grass, fine roots, and a few scraps of moss, lined with finer grass, and sometimes a little vegetable down. It is made rather deep, probably to prevent the eggs or young from being pitched out during high winds. The eggs are four or five in number, bluish-green, spotted and blotched with greenish-brown and gray. Reed Warblers are very quarrelsome little birds, and each pair take possession of some particular part of the reeds or osiers, from which they drive all intruders. This interesting bird sings incessantly through the early summer, not only in the daytime, but frequently all through the night, joining with the Sedge Warbler and the Nightingale in making the few short hours of darkness about the summer solstice resonant with melody.

Reed Warblers laying, 1st June.

The hay meadows in summer are a favourite haunt of many interesting birds, not only when the tall grass is rapidly ripening for the scythe, but when the verdant crop is laid low by the ruthless mowers, and the herbage, in long, regular swathes, is waiting for the hay-makers. How beautiful these fields look just as the crop of grass is ripening, and the summer breezes sweep across them in undulating waves, or the shadows of the passing clouds are cast upon their olive-green surface! What a wealth of life is harboured in their summertide luxuriance! Where the grass grows thickest the shy and timid Landrail skulks, or runs to and fro through the herbage in quest of food, uttering his harsh, loud cry at intervals. He is a watchful, wary creature, and is far more often heard than seen. The nest of this bird is very neat and well-made; coarse, dead grass forming the outer structure, which is lined with long fine grass. The eggs are from eight to twelve in number, buff in ground colour, spotted and blotched with reddish-brown and gray. The Whinchat is another bird of the meadows which pertinaciously keeps up its monotonous note of *u-tac, u-tac-tac-tac*, as it clings to the tall weeds and grasses. Its young are now strong upon the wing, and the little family keep together until the autumn. House Sparrows and Greenfinches perch upon the swathes of grass to pick out the seeds. Blackbirds and Thrushes love the meadow grass, and search amongst it for snails and worms, especially in early morning and at dusk, when the dew is lying thickly on the herbage. Both these species are now engaged in bringing up their second broods. The young are hatched

Landrails laying, 5th June.

Young Whinchats can fly, 10th July.

Thrushes have eggs, 2nd brood, 1st June.

in twelve days from the beginning of incubation, and are able to leave the nest and fly twelve days later still. The Skylark usually makes its nest in these fields, but its young are able to fly before the grass is mown. The Swallows glide up and down across the wavy grass, their dark metallic plumage glowing in the sunshine, bursting out every now and then into sweetest song; and the Kestrel may oft be seen beating over the hay-meadows in quest of mice. I could stand and watch the aërial movements of the Kestrel by the hour together. The beautiful grace which characterises its slow and stately flight, the activity of its movements, and the power and command it shows in falling and rising, all fill me with genuine admiration. This pretty Hawk lives almost entirely on field-mice and insects, dropping on them so softly from the sky, and bearing them off either to the nesting-place or some quiet retreat where they can be devoured in peace.

<small>Swallows in full song, 12th July.</small>

When night steals softly over the meadows, and the ghost-swift moths drone lazily up and down from stem to stem, the Nightjar wakes and leaves his daily retreat amongst the branches, or the fern, to hunt up and down in quest of food. He never fails to make his appearance at sunset, sometimes before the tree-tops have ceased to reflect the setting glory of the orb of day. You may know him by his churring cry, like the rattle of machinery. He is not shy, and will fly to and fro before you, seizing the cockchafers and big

<small>Nightjars laying, 4th June.</small>

moths that happen to be passing. The nest is on the ground, a little hollow at the foot of a tree, near the edge of the wood, or amongst the bracken and heath; the eggs, two in number, are indescribably beautiful, white in ground colour, marbled and veined with brown and gray. Another bird, most active at night, is the Landrail; his loud and monotonous *crake-crake* sounding through all the hours of darkness. From field to field he passes, calling all the time; his voice now sounding startlingly clear and distinct, anon faint and remote, as the wary bird speeds quickly through the grass, or the gentle breezes of a summer night bring it towards you or carry it away. At dusk the Barn Owl leaves the church tower, or the ivied ruin, and the Tawny Owl quits his nest in the hollow tree, both bent on capturing the mice and frogs that sport about amongst the herbage. As soon as the hay is carted from the fields, numerous birds appear upon the aftermath. Families of Tree Pipits skulk amongst the herbage, the old birds tending their broods; the first small flocks of Skylarks gather; and the Starlings in ever increasing numbers join the Rooks regularly, and feed and fly in their company. The aërial evolutions of the Starlings at this season are very interesting. You may see the birds rise in compact bunches from the grass, old and young together, and wheel and spread out and close up again in marvellous regularity, and with the greatest precision. Another bird which

Margin notes:
- Landrails cease calling, 18th August.
- Skylarks flocking, 13th August.
- Starlings congregate with Rooks, 3rd August.

never fails to make its appearance at this time of the year, is the Missel-thrush. The first and second broods quit the coppices and spinneys, and other districts where they were bred, and in small flocks repair to the grass-fields. They are excessively wary, and their rasping cries make the fields ring again. Now the Blackbirds and Thrushes frequent the gardens to feed upon the rich store of fruit, the latter birds also being very often flushed from cabbage beds, where they go to search for snails.

Missel-thrushes congregate, 24th July.

During our spring rambles we made the acquaintance of the sea-birds at several of their famous breeding-places. Now let us borrow Icarian wings, and visit the noble bird bazaars of St. Kilda, a group of small Atlantic islands some fifty miles from the most westerly of the Hebrides. The only important breeding-place in the British seas of the Fulmar Petrel is situated here, and the main colony of these birds stands unrivalled in its wonderful interest. It is situated on the face of a stupendous precipice which rises some twelve hundred feet sheer up from the restless Atlantic. The birds are now busy bringing up their young. Their single egg is laid between the 15th of May and the 5th of June, in a slight apology for a nest, and is white, rough, and chalky in texture, and smells very strongly. What pen can do justice to such a noble scene as this? The entire face of this awful cliff is one moving mass of birds. On every little grassy platform, on every ledge, and on every projection, the Fulmars cluster in one busy, active throng. When the birds are disturbed from the rocks, the scene is most impressively grand; the vast, seething cloud of birds darkening the air, and seeming as though they would descend *en masse*, and literally overwhelm us by the sheer force of their countless numbers. But little noise is heard, save the rushing sound made by the myriad wings; for the Fulmar is a very silent bird, and never utters a note of protest when its home is invaded. The Fulmar is jealously guarded by the natives of St. Kilda; it pays their rent, and supplies them with food and oil. The eggs are taken in vast quantities every spring; and in the late summer months the great event of the year at St. Kilda takes place, when the Fulmar harvest is gathered. The young birds are taken just before they are able to fly, and as many old ones as possible are knocked down or snared. For days St. Kilda is literally buried in dead Fulmars. The whole population, of some seventy souls, talk of nothing else, gather nothing else; and the strong smell from the birds and feathers is almost overpowering to a stranger. Several other very interesting birds also make these islands their home. One of them is the Manx Shearwater, a species of Petrel that breeds on one of the islets in such numbers as to literally undermine the ground, making its nest at the end of a long burrow, in which it lays a single white egg.

Manx Shearwater laying, from 13th May to 15th June.

Unlike the Fulmar, this bird is very noisy, especially at night, and when its breeding-places are intruded upon by man. Boat-loads of this curious bird are sometimes captured in a single night. The first eggs of this species are laid about the middle of May, and fresh ones may be obtained up to the middle of June. The Fork-tailed Petrel and the Stormy Petrel also breed in the St. Kilda group, making their nests in disused Puffin burrows. The former bird is only known to breed in one or two other localities in the British islands. The natives of St. Kilda will tell you that this bird is one of the earliest to arrive in spring, and one of the latest to depart in autumn. Numbers of nests are made close together; I found half-a-dozen within a patch of ground perhaps ten yards square. The nest is merely a little half-dried grass, and the single egg is white and chalky, very fragile, and spotted with reddish-brown in a zone round the larger end. The Stormy Petrel breeds in a very similar manner, and about the same time; but the single egg is frequently laid under large stones, or even in holes in walls. All these Petrels are nocturnal in their habits, and at dusk may be seen hurrying off from their holes to the sea, where they hold high carnival amongst the ocean waves, searching every crest and every hollow for their food.

Fork-tailed Petrels begin laying, 4th June.

Stormy Petrel breeding, 7th June.

The Terns are amongst the latest sea-birds to breed. The Ferne Islands are one of the most important stations where these graceful birds congregate for their summer duties. The Lesser Tern is the only species that does not breed here. Singularly enough, this charming little bird appears to prefer the coast of the mainland rather than an island for its nesting-place, and consequently it is rapidly becoming scarcer as its old accustomed haunts are encroached upon by man.

<small>Lesser Terns laying, 15th June.</small> It scrapes a little hollow in the shingle above high-water mark, and lays three or four eggs, buff in ground colour, blotched and spotted with brown and gray. There is a splendid colony of the handsome Sandwich Tern at the Ferne Islands. This bird makes a slight nest, sometimes within a few inches of the tide mark, sometimes a considerable distance from the beach, in the centre <small>Sandwich Terns have eggs, 10th June.</small> of one of the islands. The eggs are two or three in number, and remarkable for their beauty and variety, ranging from white to buff in ground colour, more or less boldly blotched and spotted with rich brown, light brown, and inky gray. The Arctic Tern and the Common Tern also breed here in abundance, the former species breeding constantly much closer to the high-water mark <small>Arctic Terns lay, 15th June.</small> than any of the rest. It never makes a nest, but deposits its two or three eggs in a hollow of the shingle, where they very closely resemble the surroundings in colour, making their discovery difficult. The Common Tern generally makes a nest, though a very slight one, and as a rule it lays its eggs some little distance from the water, among sea campion and grass. Its eggs are two or three

in number, very similar to those of the Arctic Tern, buff of various shades, spotted and blotched with brown and gray. The eggs of both these Terns are laid about the same time. It is a stirring sight to witness the anxiety of the Terns when their nests are menaced. They crowd into the air, and flutter up and down like snow-flakes, ever and anon uttering their sharp *cricking* notes. Only one brood is reared in the year, and as soon as the young can fly the pretty birds seem all impatient to set off on their ocean wanderings again.

On moor and on mountain the birds are busy bringing up their broods. The young Ring Ousels can almost fly, and the Meadow Pipits and Linnets, whose nests we visited in spring, have safely reared their families, and in many cases are engaged with a second brood. The young Grouse are rapidly reaching maturity; and on the highest mountains the broods of Ptarmigan run off and conceal themselves amongst the stones and lichens as we approach. The Plovers, and the Snipes, and Wild Ducks are all full of family cares, intensely anxious for their helpless offspring, and indulging in a hundred cunning artifices to lure us from them. Here and there we may chance to come across a young Cuckoo being fed in a most conscientious manner by its foster parents, the Meadow Pipits. Hard at work indeed are the little birds obliged to keep to feed the greedy impostor, already three or four times

^{Ptarmigan hatching, 20th June.}

^{Young Cuckoos out of nest, 24th June.}

as big as his generous providers. The nests of our little friends, the Siskins and the Twites, are packed full of hungry youngsters. On the seashore the young Gulls and Terns, like balls of brown and yellow down, are ever clamouring to their parents for sustenance.

Bird life in summer undergoes many important changes which the careful observer will not fail to notice. Most striking fact of all, especially amongst singing birds, is their loss of voice. In the middle of summer the song of the Thrush and the Blackbird is visibly on the wane; and as the days go by most of the songsters that made the early spring-tide glad with their voices will warble less and less frequently, until by the end of July their music is hushed until after the autumnal equinox. From the end of June bird music is rapidly on the wane; already we miss the glorious richness of the morning concert, and the sweet variety of the even-song. Even in the latest days of summer the Greenfinch and the Yellow Bunting sing most persistently from the shrubberies and the hedgerows; but the Tree Pipit is silent and the Whinchat's song is hushed. As a rule birds do not sing so freely during the time the young are being reared; and once that duty safely over they generally cease all attempts at music, and prepare for their annual change of dress. The following table will again help to show the state of bird music, this time during the months of summer:

AMONG THE BIRDS IN SUMMER.

Name of Species.	June, 30 ds. sang on	July, 31 days. sang on	August, 31 days. sang on
1. Missel-thrush .	Silent	Silent	Last 5 days
2. Song Thrush .	30 days	First 18 days	Silent
3. Blackbird . .	30 days	First 10 days	Silent
4. Robin . . .	30 days	First 10 days	Last 23 days
5. Wren . . .	30 days	31 days	Very irregularly
6. Hedge Sparrow	30 days	26 days	First 12 days
7. Willow Warbler	30 days	27 days, ending irregular	Last 23 days
8. Chiffchaff . .	30 days	26 days	Very irregularly
9. Whitethroat .	30 days	First 18 days	Silent
10. Blackcap . .	30 days	First 6 days	Silent
11. Redstart . .	30 days	Silent	Silent
12. Starling . .	30 days	Very irregularly	Last 23 days
13. Meadow Pipit	30 days	First 7 days	Silent
14. Tree Pipit. .	30 days	First 15 days	Silent
15. Chaffinch . .	30 days	First 10 days	Silent
16. Yellow Bunting	30 days	31 days	First 14 days
17. Greenfinch .	30 days	31 days	First 17 days
18. Skylark. . .	30 days	29 days	Silent
19. Cuckoo . .	30 days, declining	First 8 days	Silent

Some of the very earliest breeding birds may still be found engaged in nesting duties towards the close of summer. The Hedge Sparrow rears brood after brood during spring and summer right up to the period of the autumn moult; and in a similar manner the noisy restless little Wren is remarkably prolific, often being engaged in bringing up a family with the turn of the leaf. It is probable this bird rears as many as three or

_{Hedge Sparrows build- ing, 3rd July.}

_{Wrens build- ing, 29th July.}

four broods in a season. House Sparrows may almost be said to breed nine months out of the twelve, and I have even known them to do so in mild winters. The Ring Dove is another bird whose fecundity is large, and may be found nesting all through the summer and during the greater part of autumn. Yellow Buntings and Greenfinches also rear several broods, often beginning to build a new nest long after the hay crops are gathered.

<small>Greenfinch building, 4th August.</small>

The birds that breed earliest in spring are naturally the first to moult. No bird sings during this trying period of its life. Thrushes begin to drop their primaries first; and a walk through the haunts of bird life at the close of July will reveal abundant evidence in the shape of cast feathers, that the great autumn moult has commenced. Birds now become remarkably shy and retiring, and keep out of view for days together. They love to skulk amongst the densest parts of their haunts, only take flight when absolutely compelled, and are weak and sickly, feed very little, and sit and mope in the shade by the hour together. Many birds, especially Gulls, Geese, and Ducks, retire miles away from land to moult their plumage; and in some species the wing feathers fall out so rapidly that they are absolutely unable to fly until the new quills grow. With most other species the quills, the most important feathers of all, are moulted in pairs, rather slowly, so that the bird is not unable to fly at any time

<small>Thrushes begin moulting, 23rd July.</small>

during the change of plumage, although most birds fly slower, and, indeed, show little desire to go into the air at all. As soon as the change of dress has been safely accomplished, the Robin, the Wren, and the Hedge Sparrow, begin to warble just as sweetly as of old; so, too, does the Willow Wren, and a few other species; but the rule is for birds to lose their song entirely in the moult, not to regain it until love dawns anew in their little breasts the following spring. The Swallows still warble at intervals as they flit across the fields and over the water, many of the young birds making attempts at song. These birds do not moult until they reach their winter quarters in South Africa, leaving us in their worn and abraded plumage. *[Robins over moult, 18th August. Willow Wrens moulting, 8th August. Young Swallows sing, 15th July.]*

Another prominent feature in bird life during the late summer days is that many species begin to gather into flocks and to change their ground. Some of the birds which have bred on the upland moors, especially the Dunlin, depart for the mud-bound coasts. Amongst the first to be seen gregarious are the Starling, the Greenfinch, the House Sparrow, and the Twite. These gatherings are mostly of young birds, which confine themselves almost exclusively to the newly-mown meadows, the pastures, and the corn-fields. In the woods parties of Jays and Magpies may be met with, trooping along in a straggling train before the observer, and hiding themselves as soon as possible amongst the bushes. The timid *[Young Dunlins on shore, 1st August. Twites flock, 14th June.]*

Pheasants stray out of the covers on to the clover-fields and into the standing corn; whilst the coveys of Partridges rise in startled haste from the quiet corners. Down the hedgerow sides broods of Long-tailed Tits and their parents hurry along like little balls of feathers trooping in a straggling train from bush to bush. Everywhere we come across nests whose little inhabitants have flown; only a few droppings and a little scurf left behind to tell the tale of departure. Blue Titmice, and their larger congeners, the Great Titmice, are now and then seen and heard among the trees; they call little now, are difficult to see among the leaves, and their plumage has lost most of its spring beauty and brilliancy. Here and there we may come across family parties of Spotted Flycatchers. These birds love to sit upon the iron railings and wooden fences round the new haystacks, and the young ones wait patiently while their parents catch the passing insects for them. The Flycatcher is expert at his business, and rarely misses his quarry; the sharp snap of his beak, as it closes upon the poor fly, being audible at a dozen paces. The young birds may be readily distinguished from their parents by their much more spotted appearance. On the moors the Ring Ousels show an inclination to pack. As soon as ever the mountain fruits are ripe these birds congregate to feast upon them, and even wander to the gardens near the moors for a similar purpose. In the shrubberies vast numbers of young Blackbirds, in the rich brown and black dress of immaturity, may be seen; and young Robins, in the speckled dress of youth, are heard making occasional attempts at song. The Red Grouse are strong on the wing, provided the season has been a favourable one, by the 4th of August; and this species soon begins to pack after the shooting on the 12th commences. August is the great turning-point of the year among the birds, and in a hundred ways they foretell the advent of the great change of season slowly approaching. The earliest arrival among the hordes of northern waders that come southwards with the sun is the Knot. The young birds, with one or two old and possibly non-breeding birds, are the first to arrive on the mud-flats, many of them with the nestling down adhering to parts of their plumage. Young Curlews also quit their birth-places towards the end of summer, and gather on the low-lying coasts and estuaries. The Swifts, having now safely reared their broods, may often be seen feeding the young birds high up in the air. These birds are preparing for their southern flight.

Young Knots arrive, 5th August.

Young Swifts leave, 26th August.

Summer is the time when bird life sees its greatest variety, change, and activity; and the long days, and short, warm nights, we are able to spend among the birds in quiet contemplation of their little ways and secrets, may be justly marked as red-letter ones in the calendar of the naturalist.

CHAPTER III.

FEATHERED FRAUDS.

It may not be out of place to devote a chapter of this little volume to the various wiles birds display and the frauds they practise, either for their own concealment or the protection of their eggs and young. Summer is perhaps the best time of the year to study this interesting subject, for it is a season when most birds are engaged in domestic duties, especially those who exercise these deceptive arts. It is profoundly curious and interesting to note how widely Deceit is practised by the lower forms of life; and it is most probable that man acquired this objectionable trait in his character from observing similar ones among dumb creatures. Savage man is an acutely observant animal; he sees and notes every little circumstance, and is ever ready to adopt any new idea, especially if he finds that it may give him some advantage over his fellows. Grant this, and it is easy to see how a savage might, by watching a bird or an insect sham death to escape detection, or by seeing an animal engage in some cunning device by which it reaped a benefit, pursue a similar course to help himself out of a difficulty, either in getting the better of his enemies, taking an advantage of his prey, or escaping some danger. Man is a marvellous mimic, and has almost unconsciously adopted many peculiarities from creatures far below him in the scale of life; he has been anticipated by the lower animals, and even by plants, in many of his proudest achievements, and has taken hints from them in carrying out some of his most skilful handiwork. Some of the deceptions practised by insects are little short of marvellous; the arts, and tricks, and travesties followed and indulged in by birds are little, if any, less wonderful.

These peculiarities are not confined to any one class of birds; but predaceous species, and those big and strong enough to defend themselves, do not very generally descend to the practising of such deceits and wiles as their weaker and more defenceless relatives are compelled to do. We cannot imagine the plucky little Sparrowhawk descending to tricks of any kind, either in circumventing his prey, or shielding his young or himself from harm; he captures his quarry in fair fight, with impetuous dash; he defends his offspring with undaunted courage; and he dies game to the last. It is a well-known and most interesting fact that many species of insects closely resemble or mimic other and very distinct species, and it is also well known that these masqueraders derive benefit from such a strange procedure, for the insects they imitate are unpalatable to birds, who never molest them. Among birds, such instances of imitation are rare, though doubtless many remain to be discovered. Our own Cuckoo is the most familiar example of this curious kind of mimicry. Cuckoos are weak and defenceless creatures, gifted with no powerful weapons of war, and consequently they have sought protection from their greatest enemies by imitating them very closely in appearance and movements. The Cuckoo is remarkably like a Sparrowhawk, and often flies in much the same manner from tree to tree. Such a close resemblance to the Raptores serves the Cuckoo in good stead; for repeatedly the Hawks and Falcons stay their course towards him, deceived by his resemblance to one of their own kind. Small birds are equally taken in by the false pretences of the Cuckoo, and buffet him unmercifully, as they often do the Hawks and Crows.

Many birds seek to evade danger by keeping absolutely still, and making themselves look as much like the ground as possible. Any one who has travelled across desert country will be surprised how close most birds crouch to the ground, rarely taking wing until they are almost trodden upon. Birds of the Desert are almost invariably sand-coloured on the upper parts, and thus, by remaining motionless, they escape detection by their numerous enemies. I have repeatedly had Desert Larks rise from under my feet, and here and there from the ground all round me, without being able to see a bird, no matter how closely I scrutinised every foot of sand. Then the Sand Grouse, and the Coursers, and the Bustards are all guilty of this trickery in keeping close to the ground where their brown plumage harmonises so closely with the sandy soil that detection is defied. On our own mountain tops, the Ptarmigan is another noteworthy instance of this peculiar kind of fraud. You may walk right through a crouching flock of these birds, in their brown and gray livery, without seeing a single one, so closely does their plumage assimilate in colour with the moss, lichen, and pebbles amongst which they lie concealed. Again, the Snipes are remarkable in this respect, their brown and yellow-striped upper plumage resembling closely the dead blades of grass and the withered leaves; the Woodcock, especially, being beautifully adapted in this respect. And, then, how closely the plumage of the Red Grouse resembles the heather—a coincidence which saves this bird from his numerous enemies. The hen Pheasant is another instance, for at the least alarm she crouches low amongst the drifts of dead leaves, or in the dry grass, where detection is almost impossible. These are but a very few examples, but the reader will call to mind scores of others equally as startling and strange. What is, perhaps, more interesting than the actual resemblance between the plumage of these birds and their surroundings is the wonderful intelligence which prompts them to keep quiet and motionless as death during the critical moments when the least movement might lead to discovery and destruction.

Another of the tricks which some birds resort to is that of shamming death. I once caught a Landrail amongst the long grass—the bird refusing to leave the cover, which gradually got less and less as the mowing-machine went round, cutting swathe after swathe—and was as much surprised as startled at seeing it, to all intents and purposes, die in my hand. Suspecting some trick, I laid it quietly on the grass; but not a muscle moved, the eyes were closed as in death, and the legs and wings seemed to have lost all their power. There it lay for a few minutes, until I took it in my hand again, the head drooping just as though the bird were dead. Then I actually dropped it on the grass again, and retired a few paces to watch. The eye gradually opened a little, and, finally, the cunning bird sprang up and hurried away, as if nothing at all were the matter! I could not help admiring the wonderful command the bird must have had over itself—what in a man we should justly term an iron nerve under such trying circumstances. Verily birds are gifted with more intelligence, with more mental capacity, than we are ever disposed to accord to them. I have also known the Wryneck to feign death in just the same remarkable and astonishing manner.

Although scarcely coming within the province of the present subject, we can scarcely touch upon it without calling passing attention to the many devices birds employ in concealing their nests from enemies. The admirable way in which the Chaffinch tries to deceive us in building its beautiful home, or the cunning manner in which the Wren and the Dipper seek to cheat us in the discovery of their nests, are known to every field-naturalist. But birds themselves practise much fraud in shielding their eggs or helpless offspring from enemies. There are no more cunning feathered frauds than the Sandpipers, and none of these birds more tricky and deceitful than the Common Sandpiper or Summer Snipe. This little bird breeds on the shores of most of our more northern sheets of water, making its nest close to the bank of the pool, usually under the shelter of a heather tuft or tall weed. If the sitting bird is satisfied that you do not see her, she will remain on her nest until you almost tread upon the eggs, then hurriedly springing from her charge, she reels and tumbles along the ground before you, trailing her wings as if they were broken, and trying by every cunning artifice she can command to get you to run after her. You may perchance be inclined to follow the apparently wounded bird, when after going a score paces in this eccentric manner, she will rise with an exultant "weet" of joy, and fly rapidly away! One more instance may be taken from our commoner birds. Every observer of Nature is familiar with the ways of the Lapwing, when its nest or young is menaced by danger. The watchful birds rise from the moorlands and the fallows the moment you set foot upon them, and wheel and tumble in the air, uttering their plaintive cries, seeking to attract your attention to themselves. With an artfulness we cannot help admiring the Lapwings become more and more anxious the farther you get from their nests, and very often, when the eggs are lying exposed to view at your feet, the parents will assume the utmost indifference. Should the birds have young, their actions become even more demonstrative, and I have known them on such occasions tumble along the ground in anguish, and appear wounded and helpless for the few brief moments which give the nestlings time to scatter and conceal themselves among the herbage.

Nestling birds are also full of deceit and trickery; they are born with it in their nature, and practise many wiles to save themselves from harm, without tuition or experience. The young of all these Sandpipers and Game Birds and Plovers are hatched covered with down, which is striped and mottled with colour best adapted to concealment among the vegetation or shingle where they are born. The moment the more wary and watchful old birds give the warning cry, the broods of young scatter to all points of the compass, and conceal themselves wherever they can. Then they remain as quiet and still as though they had suddenly been transformed into stones, and no noise will rouse them from their assumed stupor until the mother's well understood cry assures them that all is safe once more. The first natural impulse of a young bird of almost any species is to try and conceal itself, either by creeping into a hole or squatting close upon the ground. Some young birds, such as the chicks of the Grebe or the Waterhen, dive into the water the moment harm betides, and come up far away among the reeds and rushes growing in the water, or conceal themselves in the fringe of vegetation round the bank.

I cannot leave this interesting subject of feathered frauds without giving an extract from an old note-book of mine relating to the Ringed Plover. It was in June, on the low, sandy coast of the Wash, between Wainfleet Haven and the now rapidly rising watering-place of Skegness, then a mere village, that I came across this charming little feathered favourite of mine and observed the following incident: I had known this bird bred here, in company with the Lesser Tern, for years, and had often taken its eggs from the sand; but to-day the birds had young ones only a few hours old. I found the stained and broken egg-shells lying on the shore, and the anxious movements of the old Plovers told me the young were not far off. For an hour or more I tramped up and down among the shingle without discovering a single nestling; all the time the parent Plovers were piping anxiously, and now and then flew up almost to my feet and turned rapid somersaults along the shore. Then I retired to a distance to wait and watch, and soon by the aid of a glass I saw the old birds settle and the four young ones run out from among the round pebbles and commence clamouring for food. Again I hastened to the spot. Up rose the parent birds piping into the air, and the same tricks were resorted to, but not a nestling could be seen. After ten minutes' search among the stones I found two tiny striped balls of down, with long legs tucked up beneath them, but the closest and most patient search failed to reward me with a sight of the other two nestlings. At the first alarm these little nestling Sand Plovers scattered and hid themselves among the stones, remaining as still as the pebbles themselves, until all chance of discovery had passed. It is a most interesting fact that the Ringed Plover prefers to lay its eggs on the fine sand where the small spots upon them are more in harmony with the ground than they would be among coarser shingle; but as soon as the young are hatched the old birds lead them to the rougher beach where pebbles, broken shells, and various shore *débris* afford a ready harbour of refuge in times of danger.

Such are a few of the tricks and travesties that birds play off upon their enemies. We cannot refrain when observing the cunning and deceit which wild birds practise so adroitly, from admiring the mental powers which prompt and guide these curious and wonderful movements, and the long course of experience and selection of the fittest individuals which have been necessary in bringing them to their present perfection and high degree of utility.

CHAPTER IV.

THE WAYS OF BIRDS.

To enjoy thoroughly the pleasures of the woods and fields, you must take all their wild inhabitants into your confidence, help them in their little trials, partake of their sorrows, and share their sympathies and joys. Most of us have some pet hobby which we are apt to mount and ride, when, perchance, the troubles and annoyances of life press unusually heavy upon us. Mine is Ornithology. From boyhood's very earliest days, the birds have never failed to furnish me with a constant round of enjoyment; to my feathered friends I owe a very great proportion of the pleasures I have hitherto experienced. I delight in bird company, and love to watch their ways and movements year by year, looking upon my favourites as very old and dear friends indeed. The various species of birds have each their own little ways, which you get to know and understand, and which make your rambles in their haunts assume unwonted interest. You may claim all the wild birds as your very own. No need to cage or captivate them; they are as much your own as if behind prison bars, and infinitely more interesting and instructive than if moping in captivity. The Kestrel that floats and flutters in boundless freedom over the summer woods and fields, the Sparrowhawk that shoots like a bolt along the hedgerows in the dusk of evening, are as much your own as if they languished in a cage; you may see them, study their ways, and enjoy their society, better, infinitely better, than if you were to take their liberty away. The Skylark that soars beyond our vision to the clouds will come down again to earth. Cage this sweet songster, and the very best part of all his melody is lost; he is yours just as much among the clouds as though you doomed him to captivity. He will sing to you, and follow all his accustomed habits, and perform the functions allotted to him by Nature, provided you are careful not to molest or frighten him. The loving patient study of the ways of birds leads you into their confidence, and enables you to understand their ever-varying habits and movements.

As in the human species, various emotions of birds are expressed by their voice; consequently it is of the highest importance that the naturalist should make himself familiar with their several cries and songs, so that he may understand the meaning they are intended to convey. The question naturally arises: Can birds converse? Can they communicate to each other their wants and desires through the medium of their voice? The experience of a lifetime spent among birds leads me to answer this question in the affirmative. Assuredly birds are possessed of conversational powers, and are able to convey their meaning to each other. To a casual observer this may seem impossible; birds utter a few certain stock sounds, which may or may not mean anything; but the careful student of the voices of birds well knows the almost endless inflections which their cries undergo, and correlating these with the actions they accompany, he is able to comprehend much of their meaning. The chirp of the House Sparrow, for instance, is familiar to everybody; yet how few of us are acquainted with the various sweet and sibilant sounds uttered by this bird during the act of courtship, or when numbers have gathered together in the tree-tops, just previous to retiring to rest, on some still autumn evening. Every bird utters or makes a certain sound expressive of alarm or danger; and what is rather remarkable is that this signal is as well understood by stranger species as by its own kindred. Witness how the loud pipe of the Oystercatcher, or the warning cry of the Curlew, will sound the alarm along the coast, and every bird within hearing becomes on the alert at once. The various wild birds have learnt by long experience that these peculiar notes express alarm; that they are the forerunners of danger; and their first impulse on hearing them is to get out of harm's way as quickly as possible. The bond of sympathy existing between birds of many different species is as admirable as it is true. The shrill alarm cry of the Chaffinch when its nest is threatened by an enemy attracts other birds to the scene of disaster; and I firmly believe these visits are ones of condolence, and that sympathy is expressed in a variety of twittering notes. I have often seen four or five different species of birds attracted by the alarm notes of the Song Thrush, when I have been too near its nest; and the Stormcock will set an entire coppice into uproar and tumult by its grating cries. Hundreds of similar instances might be recorded. Again, birds repeatedly express sympathy for a wounded or captive companion. I have seen Gulls and Terns of at least four species hover above a wounded Oystercatcher, and utter sounds expressive of condolence—sounds which I am certain these birds never make under ordinary circumstances; and I have had Rooks follow me high in air for half a mile from the colony, when I have been carrying one of their captive comrades. Birds, though not much attached to their eggs, have a strong affection for their young, and evidently experience the greatest joy at seeing them advance to maturity under their fostering care. Old birds will twitter to their young in a strain never heard at any other time, just as an admiring mother does to her infant; and if need be, most birds will even sacrifice their life in defending or assisting their helpless little ones. There can, however, be no doubt that this affection is only transient, and when once the objects of it are removed, it soon dies out entirely. Witness how soon birds appear to forget the grief of their loss when their young have been taken away, often preparing for another brood, the male singing just as joyfully as before the bereavement. It is well it should be so. None of us would care to contemplate the face of Nature bathed in mourning tears, or see her subjects grief-stricken and joyless long.

Most of the actions and movements of birds are accompanied by notes of some kind. Gregarious birds, in flying from place to place, twitter to each other *en route*—probably for the purpose of keeping the flock united. Migrating birds keep sounding their notes as they fly across the night sky, the leaders being answered by those in the rear, so that none of the party shall accidentally stray from the rest. Notes of triumph are uttered by successful rivals. Cries of recognition and joy are sounded when birds rejoin their mates or young. The season of courtship and love is a season of song. The mother speaks her delight in sounds understood of her species after each fragile egg is laid in the nest. Fright, distress, and anger are all expressed in certain language so plainly spoken that even man may understand it.

The patience of birds is remarkable. Note, for instance, how a bird will wait for hours, when it is conscious of being watched, before it will venture near its nest. Or, observe the stolid patience of the Heron fishing by the water-side, waiting for many a weary hour in hope of food. Their perseverance is none the less extraordinary. Time after time will the House Martins begin to make a new nest as soon as the old one is destroyed. Then, how closely the Woodpeckers will dig away at the bark for insects; or, with what pertinacity birds return year after year to old familiar haunts, in spite of incessant persecution and disturbance.

A study of the mental qualities of birds is equally interesting. Birds are unquestionably gifted with extensive powers of reason, and innumerable instances might be given in support of such being the case. It is, of course, manifestly unjust to compare the reasoning capacities of birds with those of highly civilised men, for then the difference in degree is vast; but this inequality becomes exceedingly less acute when the reason of savage man is substituted for comparison. Birds, like savages, are gifted with amazing powers of perception, with deeply-rooted affections, with likes and dislikes, with acute aptitude for observation, and with a considerable taste for the beautiful, both in sight and in sound. Some or other of these various mental qualities are constantly being called into practice, and the observer will have abundant evidence of their existence if he watches the ways of birds from day to day.

The habits of birds assume a newer and a

higher interest when we study them in conjunction with the mental attributes of the birds themselves. What vast variety we find, for instance, in the degree of sociability which exists in certain species. Some birds are regular hermits, living lives of solitary seclusion, only seeking the society of their own species during the period of reproduction. Others are just as gregarious in their instincts, banding together, and performing all their functions in the company of their kind; many not only join into communities with their own species, but are exceedingly social, and mix freely with other and often very distantly related birds. Some birds are companionable or gregarious at one season and solitary and unsociable at others; some live in pairs all the year; others separate as soon as the brood is reared. Many curious facts may be observed relating to the songs of birds. Some birds are so pugnacious when engaged with their music that they will scarcely admit of a rival within hearing; whilst others delight to sit and warble, one against the other, in perfect peace and unity. A few birds sing all the year round, the moulting season excepted; others warble during the season of courtship and love alone; whilst some only at rare and long intervals indulge in song at all; as, for instance, the Brown Flycatcher, the Yellow Wagtail, and the House Sparrow.

Now a few words on the mating of birds. The rule is for birds to pair with fresh mates either for every brood, or to continue in each other's company for the entire season in such cases where more than one family is reared. Other birds pair for life; the union, once formed, never being dissolved except by death; whilst in a few exceptional cases the mating is more or less promiscuous, as in our Common Cuckoo, although, by the way, I am by no means convinced that this species does not pair for the season. Lastly, we have the polygamous birds. This peculiar habit chiefly prevails among the Gallinaceous order of birds, familiar instances being the Black Grouse and the Pheasant. In some species the affection existing between a mated pair is very deep indeed, birds often pining away over the loss of a partner; whilst in others little concern is manifested, and a widowed bird has even been known to get a new mate in a few hours.

Thus, then, we see that the passions of love and jealousy, the joys, delights, and anxieties of paternity and maternity, the affection and faithfulness between male and female are all enjoyed and expressed by birds in many different ways. They are capable of showing sympathy with distressed companions, and of combining their strength for the common good in driving off an enemy. In selecting sites for their nests, and in building their charming homes, they display an amount of intelligence neither equalled nor excelled by any other class of animals; and in their love for sweet songs, and their tastes for the beautiful, they are no mean rivals of man himself. There is much in the ways of birds not yet understood by man. We are too prone to credit them with a far less amount of mental capacity than is justly their due, and to blind ourselves to their marvellous and varied accomplishments, simply because we do not know sufficient about their ways to comprehend their meaning. The fault lies with us. Too long have we looked upon birds as so much material for the building up and creating a mass of "scientific" literature which is as dry and uninteresting as it is useless. Too long have we been content to think that the history of a bird begins and ends with a study of its dried skin. When we replace the Life behind the feathers with a stuffing of cotton wool, we take away the most charming part of the whole story which the bird has to tell. A new era is dawning. The dead birds have had their day, and naturalists are beginning to wake up to the fact that the living birds are infinitely more interesting, more wonderful, and more beautiful.

The various little ways of birds prove undoubtedly that our feathered friends do not pass the automatic and monotonous existence we are so apt to presume they do, and that their little lives are brightened and made happy by emotions, and passions, and desires, which conduce enormously to the pleasures of their own being, as well as to those of ourselves in studying and observing them.

CHAPTER V.

AMONG THE WHEAT.

With all our nineteenth-century knowledge, it is cause for surprise that man has learnt to appreciate so little the usefulness of many dumb creatures. Take birds alone for an example. I have no hesitation in saying that, in the present condition of life upon our planet, if it were not for birds many kinds of vegetation could not exist, and the practice of agriculture would be almost impossible, owing to the growth of noxious weeds and the ever increasing abundance of insect life. The vast and constant services rendered by birds to man are but little understood, and even less are they appreciated. Man, in his supreme ignorance, kills and persecutes his feathered friends on every available occasion, and consequently insects and weeds cause much needless destruction among agricultural and horticultural produce. We cannot do better, as we wander through the corn-fields this pleasant summer evening, than to devote an hour or so to the study of this important subject, from the husbandman's as well as from the birds' point of view.

We could not introduce this important matter of the birds' usefulness to man more opportunely than during a stroll through the wheat-fields. Birds of many species abound in them as soon as the grain begins to turn colour—some for plunder, others for a more useful purpose. As we wander along the hedgerow-side, flock after flock of greedy Sparrows rise from the corn; Finches of various kinds fly up here and there, and many other birds from time to time leave the waving grain.

It will, perhaps, be most convenient to take in succession the various species of birds which are regarded by farmers and gardeners as injurious to their crops, and weigh impartially the evidence against them and that in their favour. The first that we will notice is the House Sparrow. By the majority of farmers all small and dull-coloured birds are classed as "Sparrows," and the gun deals indiscriminate destruction to every feathered creature which may chance to bear the faintest resemblance to that cunning, impudent bird. Rightly or wrongly, the House Sparrow has got a bad name; that, as with a dog, is all-sufficient condemnation, and I am afraid this bird has few friends. But, kept in proper bounds, the Sparrow is of great service to man, and makes ample amends for his misdeeds in summer and autumn during the remainder of the year. It is the Sparrow's artificial conditions of existence which conduce to his abnormal increase, and which render him a pest rather than a benefit. By killing off his natural enemies, the Hawks, we remove one of the greatest checks to his increase; and by affording him such shelter in towns and villages, and such a regular and constant supply of food, we keep up to its full strength the army of Sparrows which come down every summer to the corn-fields and gardens surrounding them. Great as is the Sparrow's increase, I do not think farmers and gardeners would suffer one tithe of the loss from this bird if they would protect the Sparrowhawk. As things are now, the Sparrow has undisturbed possession of the fields—not a Hawk is left in many districts, and the criminals flourish because the police are withdrawn. Half-a-dozen Hawks would keep the Sparrows in bounds, even in the worst infected districts, and save many an acre of grain from the serious threshing it receives from these troublesome birds. The House Sparrow has greater powers of adapting itself to circumstances than any other bird; herein lies the secret of its enormous increase. Large villages and towns are the chief nurseries of the House Sparrow, where, unmolested, it is allowed to bring up brood after brood for nearly nine months every year, the vast numbers of young birds descending in due course upon the nearest fields of grain. It has been said that the House Sparrow is increased in numbers by migrants, but such is not the case; and the vast flocks we see upon the fields are caused merely by local movements—from villages and towns to the country districts. The damage these enormous flocks of Sparrows do to grain crops alone is very considerable—at the lowest possible estimate for the whole country, say upwards of a quarter of a million sterling every year! This is a very serious item in the agricultural statistics of the United Kingdom, and one which is almost invariably overlooked; yet the fact remains. Before we discuss remedies, let us look on the other side of the picture.

So far as my own experience of the House Sparrow extends in this matter, and it is based on much careful dissection, I am strongly of opinion that the bird's greatest use to man rests in its eating the seeds of various weeds. Sparrows feed greedily on the seeds of all the commoner weeds, such as dock, chickweed, dandelion, bindweed, charlock, etc.; and in this respect alone pretty well compensate the farmer for the damage to his grain, although the benefit he derives is not so apparent to him as his loss. Throughout the entire year the customary food of the House Sparrow is grain and seeds. For three months out of the twelve—July, August, and September —the Sparrow has the corn-fields at his mercy, but during the other nine months the seeds of weeds are his staple support. Of course this only applies to the country Sparrows, the town birds live on anything and everything they can pick up in the streets and near houses; and as soon as grain becomes scarce the fields are deserted by them. During the spring and summer the

Sparrow eats many insects and their larvæ, but appears always to evince a partiality for seeds. Its young, however, from the time they are hatched to the period of their flight from the nest —about a fortnight—are fed almost entirely on insects and caterpillars; and as the Sparrow rears many broods in the year, the amount of insect life destroyed must be very considerable. It should, however, be stated that the later broods are often partly fed on the young and milky grain.

The questions therefore naturally arise: Is the Sparrow worth his keep? Does the good he effects compensate and balance the harm? In this practical age when every item of cost of production must be considered, are the farmers justified in waging a war of extermination against this bird? Even we ornithologists, and therefore champions and lovers of all birds, cannot blind ourselves to the fact that the Sparrow is undoubtedly injurious to grain crops. The bird is too numerous. It is allowed from a variety of causes to multiply too quickly; and last, and most important fact of all, its enemies, the natural checks to such abnormal increase, are not allowed to live. I would make it an offence to destroy a Sparrowhawk, or any other bird of prey in agricultural districts, in spite of the protests of game preservers. These Hawks play an important part in agricultural economy, and should be preserved. There is no fear of our being overrun with Hawks, their numbers would be regulated by the supply of food. It is no use putting a price on the head of the Sparrow—it is a cruel and debasing practice, and, at best, an unsatisfactory one. Encourage the Hawks, and the Sparrows would just as certainly decrease. There is no need for their complete extermination, for kept in proper bounds the House Sparrow is undoubtedly a staunch and valuable friend to the farmer. If farmers are not intelligent enough to look after their own interests, then the Legislature must do so for them. The matter is of national moment. The farmer must be taught to recognise his friends, and the small game-preserving interest must not be allowed to stand in the way of the vitally important grain-growing industry. The Hawks and other predaceous birds must live, and the vast surplus Sparrow population must die!

The Sparrow is nothing near so destructive in the garden as on the farm. It may have an inordinate love for peas, currants, and gooseberries, and in seed-time scratch up the beds, but the damage is small, for the birds do not congregate in any large numbers in such localities. Sparrows are also rather destructive to flowers, and often take delight in pulling to pieces crocuses, pansies, dahlias, and other blooms. They also do some damage to the blossoms of fruit trees, but their depredations in this respect are easily controlled and frustrated.

But the Sparrow is not the only feeder on grain. There are one or two more of our

British Finches which regularly frequent the corn-fields during the summer. Most noteworthy of these are the Greenfinch, the Common Bunting, and the Yellowhammer, the former especially loving to congregate in flocks upon the corn as soon as the grass is turned into hay. The damage done by these birds, however, is only trifling; not that they are any less voracious than the Sparrow, but simply because they do not occur in such vast numbers. Again, these Finches and Buntings do far more good than ever man can estimate in ridding the fields of weeds; whilst all through the spring and during a greater part of the summer, their diet is more or less of an insectivorous nature. There are many other birds which become more or less troublesome to the farmer and gardener in seed-time, but these species invariably leave the ripening crops alone. Amongst them may be noted the Chaffinch and the Skylark; both of which never fail to make their appearance on the beds and newly-sown fields, but the harm they do is comparatively small. It must also be remembered that these birds are great seed-eaters, consuming incredible quantities of the seeds of plants injurious to husbandry; and during a great part of the year they not only subsist on insects, but bring up their young on similar food. Then among the farmer's friends we must include the army of smaller Finches, such as Redpoles, Linnets, Twites, and Goldfinches. None of these birds eat corn; all through the year they subsist on the seeds of weeds and on insects. How useful, for instance, is the Goldfinch. This charming bird prefers the seeds of docks and thistles to any other food, and these are two of the most stubborn and troublesome of weeds that the farmer has to contend with; then, again, the Bullfinch searches along the weedy slovenly hedgerows for the seeds of many noxious plants, therefore let this be remembered in his favour during the time he gets into mischief among the buds of the fruit trees; whilst the Brambling in flocks rids stubbles and pastures of weeds innumerable. These Finches, therefore, are amongst the farmer's best friends; yet too often does he shoot them down as "Sparrows," ignorant of their usefulness, and blind to their good offices on his pastures.

All the birds we have hitherto noticed have belonged to the sub-family Fringillinæ, or birds with a hard, conical beak, adapted to crush corn and seeds, and whose diet is principally granivorous; we shall now glance at the various soft-billed birds that are often found among the wheat. One of the most regular visitors to the corn-fields is the Whitethroat, another is the Willow Wren, a third is the Tree Pipit, a fourth is the Hedge Accentor. Now, ninety-nine farmers out of every hundred firmly believe that every bird they see in or near a wheat-field is there for no good purpose; but all the little birds I have just mentioned go there repeatedly for insects, and, with the exception of the Tree Pipit, for nothing else. This latter bird sometimes eats the young milky corn, but the damage done is practically nothing. The farmer's hedges and fields, all the summer-time, are full of these and other soft-billed birds, busy all the hours of daylight that they stay in this country in search of insects; ridding vegetation of pests which, if left unchecked, would work ruin and devastation on every side. These birds do no harm whatever to man or his property, and take no share of the harvest, which they so largely help to protect, as their just reward.

There are, however, some birds that fly about the corn-fields which the farmer looks upon as his friends. Would that the same welcome were extended to the rest! These are the Swallows, Martins, and Swifts. The farmer never molests them; he often forbids his servants to harm them, and suffers them to use his barns and outhouses for nesting-places. Yet these Swallows and Swifts are only doing high up in the air what scores of other little creatures are doing among the leaves — searching ceaselessly for insects. Another bird few farmers care to molest is the Wagtail. Its usefulness is apparent to them, as they see it run daintily along the furrows at the heels of the ploughman, or attend the labourers in the turnip-fields. In these latter situations the Wagtail is of inestimable service, its chief food being the dreaded " fly." None the less useful are the Pipits, although they are not so well known, and often confused with Sparrows.

A few words now about Rooks and Starlings. There are very few farmers who do not give the Rook a bad character, and show their appreciation of its services by hanging dozens of these sable birds over their fields in seed-time, as a warning and a scarecrow to the rest. The Rook is one of the agriculturist's best friends, and, with the exception of a little pilfering in sowing-time, is constantly ridding the ground of creatures injurious to vegetation. There are few greater pests than the wireworm, and the Rook is its greatest enemy. I have seen the crop of this bird extended to its utmost limit with wireworms, whilst grubs and larvæ of all kinds are eagerly sought. In like manner the Starling is a valuable assistant to the farmer, never meddling with his crops at any stage of their growth. There are few more harmless birds; yet the poor Starling is persecuted, accused of stealing Pigeons' eggs—owing to its frequenting the Dovecote for breeding purposes—and too often shot down in autumn for the sake of a handful of elderberries. A word of commendation should also be given to the poor persecuted Nightjar, the Cuckoo, the Landrail, and the Magpie—all birds leading harmless and useful lives in the meadows and pastures.

As the gloom of evening settles over the corn, the Owls leave their daily haunts in the distant woods and farm-buildings, and hunt about in quest of their prey. Farmers, and more especially game preservers, persecute these birds incessantly, without giving a thought to their usefulness. The corn-fields swarm with mice and rats, so likewise do the barns and outbuildings. The usual food of the Barn Owl and the Wood Owl is mice and rats, and the number of these little animals caught among the wheat is incredible. Softly the useful birds pounce down again and again on their prey, ridding the fields of their destructive pests, claiming no reward from man, asking only to be let alone. Every year these useful birds are becoming scarcer; the ignorant intolerance of farmers and keepers is slowly but surely working their extinction, and the mice and rats will soon have things much their own way.

Here, then, we will leave the birds to the farmer's careful consideration. Let him watch the habits of these feathered policemen, and convince himself of their usefulness. Let all lovers of the feathered tribe plead their cause with the farmer and the gardener; and, if this will not do, let us have these birds placed more effectually under the protection of the law. This question is one of vital importance to the industry of agriculture, and, all sentiment aside, is one which irresistibly appeals to every lover of birds.

CALENDAR FOR SUMMER.

Species.	June.	July.	August.
Golden Eagle	Breeding	Young wander South	Moulting
White-tailed Eagle	,,	,,	,,
Osprey	,,	,,	,,
Peregrine Falcon	Young leave parents	,,	,,
Hobby	Breeding	Moulting	,,
Merlin	,,	,,	,,
Kestrel	,,	,,	,,
Honey Buzzard	,,	Breeding	,,
Marsh Harrier	,,	,,	,,
Hen Harrier	,,	,,	,,
Montagu's Harrier	,,	,,	,,
Sparrowhawk	,,	,,	,,
Common Buzzard	,,	Moulting	,,
Barn Owl	,,	,,	,,
Long-eared Owl	,,	,,	,,
Tawny Owl	,,	,,	,,
Missel-thrush	2nd brood	Begin to flock	,,
Song Thrush	,,	Moulting	,,
Blackbird	,,	,,	,,
Ring Ousel	Breeding	,,	Congregating
Dipper	,,	,,	Moulting
Wheatear	,,	,,	Moulting and migrating (young)
Whinchat	,,	Young and parents in fields	Moulting
Stonechat	,,	Begin to moult	,,
Redstart	,,	,,	,,
Robin	,,	Ceases to sing	,,
Nightingale	,,	,,	,,

CALENDAR FOR SUMMER.

Species.	June.	July.	August.
Whitethroats	Breeding	Cease to sing	Moulting
Blackcap	,,	Ceases to sing	,,
Garden Warbler	,,	2nd brood	,,
Reed Warbler	,,	Breeding	,,
Sedge Warbler	,,	Ceases to sing	,,
Grasshopper Warbler	,,	2nd brood	,,
Chiffchaff	,,	Engaged with young	,,
Willow Wren	,,	Moulting	,,
Wood Wren	,,	Breeding	,,
Goldcrest	,,	Congregating	,,
Great Titmouse	,,	2nd brood	,,
Blue Titmouse	,,	,,	,,
Coal Titmouse	,,	In parties	,,
Marsh Titmouse	,,	,,	,,
Hedge Accentor	,,	3rd brood	,,
Wren	,,	,,	,,
St. Kilda Wren	,,	—	,,
Creeper	,,	2nd brood	,,
Nuthatch	,,	In families	,,
Raven	Young leave parents	Moulting	,,
Carrion Crow	Breeding	Congregating	,,
Hooded Crow	,,	Wandering in broods	,,
Rook	Unite into flocks	On pasture lands	,,
Jackdaw	Breeding	,,	,,
Magpie	,,	Seen in parties	,,
Jay	,,	In parties	,,
Red-backed Shrike	,,	Breeding	,,
Starling	2nd brood	Moulting	In flocks with Rooks
Goldfinch	Breeding	Breeding	Moulting
Siskin	,,	2nd brood	,,

Species.	June.	July.	August.
Greenfinch . . .	Breeding	Breeding	Loses song in moult
Hawfinch	,,	,,	Moulting
House Sparrow . .	2nd brood	,,	,,
Tree Sparrow . .	Breeding	,,	,,
Chaffinch	,,	Moulting	Moulting and silent
Linnet	2nd brood	,,	,,
Lesser Redpole . .	Breeding	Breeding	Young in flocks
Twite	,,	In flocks	Moulting
Bullfinch	,,	Breeding	,,
Corn Bunting . .	,,	,,	,,
Yellow Bunting . .	,,	,,	Loses its song
Cirl Bunting . . .	,,	,,	Moulting
Reed Bunting . .	,,	In attendance with young	,,
Pied Wagtail . . .	,,	,,	,,
Gray Wagtail . . .	,,	,,	,,
Yellow Wagtail . .	,,	,,	,,
Meadow Pipit . .	,,	Congregating	,,
Tree Pipit . . .	,,	In attendance with young	,,
Rock Pipit . . .	,,	,,	,,
Skylark	,,	,,	In flocks
Wood Lark . . .	2nd brood	,,	Moulting
Swallow	Breeding	Breeding	Young gather into flocks
Martin	,,	,,	Young congregate
Sand Martin . . .	,,	,,	,,
Swift	,,	,,	,,
Goatsucker . . .	,,	,,	Moulting
Kingfisher . . .	,,	Moulting	,,
Green Woodpecker.	,,	,,	,,
Great Spotted Woodpecker	,,	,,	,,

CALENDAR FOR SUMMER.

Species.	June.	July.	August.
Lesser Spotted Woodpecker	Breeding	Moulting	Moulting
Wryneck	,,	,,	,,
Cuckoo	,,	Ceases to sing	Old birds leave
Ring Dove . . .	2nd brood	Breeding	Moulting
Stock Dove . . .	,,	,,	,,
Rock Dove . . .	,,	,,	,,
Turtle Dove . . .	Breeding	,,	Congregating
Ptarmigan	,,	Gather into flocks	Moulting
Red Grouse . . .	,,	Moulting	Begin to pack
Black Grouse . .	,,	In broods	Moulting
Capercaillie . . .	,,	,,	,,
Pheasant	,,	Breeding	,,
Partridge	,,	In broods	,,
Quail	,,	Breeding	,,
Heron	Young wander far from home . .	Non-gregarious	,,
Corn Crake . . .	Breeding	Breeding	Ceases to call
Spotted Crake . .	,,	In parties	Moulting
Water Rail . . .	,,	2nd brood	,,
Moorhen	,,	,,	,,
Coot	,,	In family parties	,,
Stone Curlew . . .	,,	In fields with young	,,
Oystercatcher . .	,,	Breeding	In parties
Ringed Plover . .	,,	,,	,,
Golden Plover . .	,,	,,	,,
Lapwing	,,	Congregating	Moulting
Red-necked Phalarope	,,	Breeding	,,
Curlew	,,	Congregating	Appear on coasts
Whimbrel	,,	,,	,, (young)

Species.	June.	July.	August.
Common Sandpiper	Breeding	Breeding	In parties
Redshank	,,	Congregating	Moulting
Greenshank . . .	,,	,,	,,
Knot	—	—	Young birds arrive
Dunlin	Breeding	Congregating	Young on mud-flats
Sanderling . . .	—	—	,,
Woodcock . . .	Breeding	Breeding	Moulting
Snipe	,,	,,	,,
Sandwich Tern . .	,,	,,	Begin to leave stations
Common Tern . .	,,	,,	,,
Arctic Tern . . .	,,	,,	,,
Lesser Tern . . .	,,	,,	,,
Black-headed Gull .	,,	Young and old leave gulleries	Moulting
Common Gull . .	,,	Breeding	Begin to leave stations and moult
Lesser Black-backed Gull	,,	,,	,,
Great Black-backed Gull	,,	,,	Young congregate
Herring Gull . . .	,,	,,	,,
Kittiwake	,,	,,	Leave breeding places
Great Skua . . .	,,	,,	Wander southwards
Richardson's Skua .	,,	,,	,,
Puffin	,,	,,	Leave breeding places
Razorbill	,,	,,	,,
Black Guillemot . .	,,	Busy rearing young	Leave stations

CALENDAR FOR SUMMER.

Species.	June.	July.	August.
Common Guillemot	Breeding	Busy rearing young	Leave stations
Red-throated Diver.	,,	Young able to fly	Draw southwards
Black-throated Diver	,,	,,	,,
Great Crested Grebe	,,	,,	Moulting
Little Grebe . . .	,,	Young leave birthplace	2nd or 3rd broods
Manx Shearwater .	,,	Breeding	Wander out to sea
Fulmar.	,,	,,	,,
Stormy Petrel . .	,,	,,	,,
Fork-tailed Petrel .	,,	,,	,,
Gray-lag Goose . .	,,	Moulting	Congregating
Sheldrake	,,	Breeding	In parties
Wigeon	,,	In broods with female	Moulting
Teal	,,	,,	,,
Garganey	,,	,,	,,
Shoveller	,,	,,	,,
Mallard	,,	,,	,,
Pochard	,,	,,	,,
Tufted Duck. . .	,,	,,	,,
Eider Duck . . .	,,	Breeding	In broods
Red-breasted Merganser	,,	,,	,,
Gannet.	,,	Busy rearing young	Begin to leave stations
Cormorant . . .	,,	,,	,,
Shag	,,	,,	,,

Part III.—Autumn.

CHAPTER I.

THE BEAUTIES OF THE AUTUMN.

Hence from the busy joy-resounding fields
In cheerful error, let us tread the maze
Of Autumn, unconfin'd; and taste reviv'd
The breath of orchard big with bending fruit.

SUMMER is past! One by one its beauties have been displayed; the pageant is gone by; and before the last strains of its attendant music have died away, among falling flowers and vanishing life in countless forms, the earliest signs of Autumn's advent steal slowly over the face of Nature. A beautiful calm is settling over all things; tired Nature, after its spring and summer revels, is sinking gradually into that torpor which precedes decay; the year has passed the meridian of its splendour, and has now but to decline and die. The birth of the year, and the full time of its splendour, are perhaps no more beautiful in their several aspects than is the season of its decline and death. Autumn is the grand time of

AUTUMN.

fruition—the period when the year's increase is ripe unto the harvest; and a season when the northern world is gilded with all those lovely tints and hues, which are none the less beautiful because they are the heralds of death and decay. For the naturalist, the period of the year's decline is fraught with interest—he may then watch the habits and movements of all animated Nature, preparing for the coming winter, just as he already has observed the glories of their birth and youth in spring, and the wonders of their developing maturity in summer. Already the once sober green leaves on trees and hedgerows are rapidly changing into the browns, and yellows, and purples, which proclaim the glory of their fall. Soon the landscapes will glow in a hundred blazing tints, some blended beautifully together, others standing out in brilliant relief, like seas of russet and gold, as the sun gleams in mellow radiance athwart their broad expanse. Autumn is more apt to fill a contemplative mind with sorrow than any other season; yet Nature knows no sadness in all this quiet decay; each leaf and each bloom has completed its mission, and not one of them will fall in vain!

Our rambles through the fields and woods reveal important changes. We miss the music of the birds that made spring and early summer so beautiful; we note the absence of much of the bustle and excitement characteristic of the country during the love season of our feathered favourites;

the great symbol of autumn is a sense of rest and quietness among the animal kingdom. Silent are the woods and copses now; silent, save for the rustling of the many-tinted leaves, as they fall fitfully from the branches overhead; silent, save for the occasional cry of the Blackbird and the Robin, or the garrulous little Titmice and Goldcrests high up in the painted trees. Most of our common birds are either moulting, or scarcely yet recovered from that tedious process of changing their feathers—hence this silence and the deserted aspect of the woods. The movements of birds form one of the most imposing beauties of autumn. On every side we see signs of the vast feathered exodus. The air is full of the Swallow tribe—in fluttering hosts the little birds course to and fro, full of the excitement of setting out on their long journey to the southern shores of Africa. As autumn advances the birds complete their moult, and then we see the migrants all restless and uneasy to be gone, flying about the bushes and trees, as anxious now to quit their breeding-places as they were eager to reach them in the spring. Birds now appear to be in the very height of their enjoyment. Family cares are over; the sickly period of moulting is safely passed; food is still plentiful; the autumnal air is mild and balmy; the sunshine is yet warm and genial—all now seem given up entirely to the pleasures of existence. And yet our little feathered friends are surrounded by many perils, many dangers; not only they who flit about so happily here, but their congeners, the army of migrants, that have the long journey south before them. Hungry Hawks are ever on the look-out for prey; marauding weasels and cats watch incessantly for the weakly and unwary.

In spite, however, of all we hear about the "struggle for existence," and the "battle of life," there is abundant evidence to prove that this law of Nature entails no unhappiness upon dumb creatures. On the bending spray the Robin may be singing loud and long for very joy; like a whirlwind the fierce Sparrowhawk swoops past the brambles, and his talons meet in the quivering body of the little songster. All, however, was so sudden, so unexpected, that the warbling Robin had no time for sorrow; his misery was short, and his death almost a painless one. And so it is throughout the realms of organic life. The strong may prey upon the weak, and the powerful strive to thrust aside creatures more helpless than themselves in this universal conflict, yet the method of the warfare is attended by no unhappiness. I venture to say that birds and animals, insects and fish enjoy their life infinitely more fully than mankind. Dumb creatures are troubled with no thoughts for the future, and soon—very soon forget the past; they live in and extract as much pleasure as possible from the present. The birds sing and fly about, call to each other, feed and pursue their various habits with no thought or knowledge of the Hawk or the cat which is seeking to destroy them; the insects dance about in the sunshine, or flit to and fro in the dusk of evening, heedless of the feeding Swallows, Goatsuckers, and bats; the lambs skip about on the hillsides and the leverets in the corn, having no fear or dread of the Eagle's fatal stoop; the tiny fishes in the stream swim to and fro in their gladness, and even jump out of the water in exuberance of spirit, all unconscious of the voracious pike, the otter, or the Kingfisher, which may be about to devour them. To infer that Nature is fraught with misery and unhappiness because the fundamental condition of its existence is one of battle and murder, is a very wrong assumption, and one which the observation of wild life in its natural haunts will quickly dispel. The mortality among organic life is enormous—millions of creatures are born that do not survive their infancy; millions more perish in yet earlier stages of existence.

Among birds, for instance, which have the power of increasing rapidly, of the young, hatched one season, but a very trifling percentage survive until the next. The vegetable kingdom is under the influence of the same great law—in many cases not one seedling in ten millions surviving in the struggle for existence. This noble oak tree before us, for a thousand years or more has regularly produced crop after crop of acorns, yet only one sapling yonder in the hedge survives! Even with animals that multiply exceedingly slow, such as the elephant for example, a very few years, comparatively speaking, would suffice to stock the entire world, if each individual born could live and propagate its kind. The few organisms that do survive the perils and the dangers of existence are invariably those best adapted to the conditions of their life, and the ones most fitted to enjoy and propagate it. Thus, then, we see that this eternal warfare and this enormous mortality is an essential condition of existence, and only tends to exalt the life which is left.

Perhaps at no other time of the year is Nature's lavishness of production so apparent as during the autumn months. Take birds as an instance. Flock after flock and family after family of these creatures fill the woods and fields, the moorlands and the shore. The majority of these feathered hosts are young birds, the produce of the spring and summer. When we see them all so full of life and activity now, we cannot refrain from musing over the fate that awaits them. Ere the spring-time dawns again, by far the greater number of these young birds will be dead; they will fall victims to the countless perils of their existence, some time during the coming winter. Only a few favoured ones will survive—the fleet of wing, the strong of constitution, the wary, and those best adapted in every way to compete in the struggle for life. Nature is inexorably severe; she abhors the ailing and the weakly, these are the first to perish. By these means that standard of perfection is maintained which is requisite to keep species in existence; it prevents their deterioration and keeps all living things in harmony with the ever changing conditions of their life.

But the autumn draws on apace. Day by day we miss the birds of summer; they are leaving us in thousands with every favourable wind. Other birds pour in from higher latitudes—everywhere throughout the northern hemisphere the departure of the summer birds is the unfailing sign that winter is approaching. We miss various animals from their accustomed haunts as they retire into their "winter quarters" to lie dormant during the cold season. The bats no longer flit about at dusk round the elm trees on the common; the dormice have made themselves snug nests among the hazel bushes; the hedgehogs have retired till warmer days return. The trees are rapidly losing their brilliant-tinted foliage, the aroma from the carpet of fallen leaves being exquisitely sweet. The moorlands are all aglow with purple bloom, and here and there the mountain-ash and service tree tempt the Thrush tribe with their clusters of scarlet berries. There is a beautiful and indescribable charm about the woods and groves at this season, which they possess at no other time of the year. And then how delightful are the clear, fresh autumn mornings; how they tempt us into the fields and lanes to watch the movements of animated Nature; whilst at eventide, as the full, yellow moon rises so solemnly above the trees, the habits of birds just previous to seeking their roosting-places, are fascinating and interesting beyond all description. The broad brown stubbles which crackle sharply under our feet as we step across them, and the fields of turnips dripping wet with dew, are chosen haunts of wild life, full of charm for the naturalist; whilst the grass meadows and the brookside will reveal a rich variety of stirring incident.

At all seasons of the year birds congregate where food is most abundant. In spring and summer food supplies are more dispersed than in autumn and winter; consequently, during the latter seasons, the naturalist must pay considerable attention to this matter, otherwise he will miss much that is interesting. In autumn, wild fruits and berries are the great attraction for vast numbers of birds. Let the observer then repair to the trees and shrubs which bear these fruits, and he may be certain of finding birds in abundance. So long as the stubbles remain unploughed, they are the haunt of many granivorous birds; the fields of "seeds" or stubbles which have been sown down with clover, are an unfailing attraction for the smaller Finches. To the beech woods numbers of birds are drawn by the dropping "mast," and the oak trees are frequented for the acorns.

The beauties of the autumn! Everywhere the magic wand of autumn is changing the aspect of all things. How delightful it is to be out in the wild woods and watch this grand change slowly take place. To breathe the bracing air, pure and fresh as only the country knows it; to see the glorious tints deepen and intensify over the woods; to watch the bracken and the ferns and brambles turn into gold and brown and purple; to observe the habits of the birds, and beasts, and insects; to hold communion with Nature in the gloaming; to see the fog-banks float across the dew-soaked meadows, and melt away before the rising sun; to contemplate the maze of gossamer, spangled and beaded with dew-drops, which yokes together in fairy net-work every twig, and leaf, and grass-blade; to listen to the mellow music of the few autumnal songsters; to admire the clustering wild fruits and berries, and the painted fungi; to smell the fragrance arising from the damp, fallen leaves at morning and evening; to gaze at the brilliant "harvest" moon peacefully beaming over the sleeping fields; to muse upon the endless themes which Autumn ever inspires in a contemplative mind—all, all these are pleasures indeed which make the autumn months bright and beautiful, and surround them with a charm as bewitching as it is rare and indescribable!

Of all the seasons of the year I love the autumn best; it is my greatest delight to watch the beauties of the year's decline, seeing in them, not the signs of decay and death, but a brilliant promise and a glorious hope of new generations of life soon to rise, Phœnix-like, from the ashes of the old!

CHAPTER II.

AMONG THE BIRDS IN AUTUMN.

THE last days of summer, and the first of autumn, are marked by a striking quietness amongst bird life. The music of nearly every species of bird is hushed; most birds are more skulking than usual; many disappear to all but the keenest scrutiny; and only an occasional chirp betrays their whereabouts. Many birds are still moulting; but, as the autumn passes away, we gradually see the apparently deserted woods and fields filled once more with their feathered inhabitants —songless, to remain so until the following spring. To this rule, however, there are one or two exceptions. Undoubtedly, the homely little Robin is the most prominent songster in autumn. His sweet strains lend life to the woods and hedgerows, and the shrubberies and gardens are made glad by his music, just as the Michaelmas daisies and chrysanthemums give a last touch of welcome colour to the flower-beds. Every one knows the Robin, almost every one is familiar with his rich, sweet melody. There is something

Robins in full song, 1: September.

very plaintive about the Robin's autumnal song—a melancholy sadness that seems in harmony with falling leaves, damp, decaying vegetation, and bare boughs. As an autumn songster, the Wren is sure to attract constant notice, his song being renewed after the moult, and attaining all its loud and varied beauty as the mellow days of the Indian summer gild the waning year. The Song Thrush may sing a little now and then in autumn, and the Blackbird warbles even less frequently still; but the mellow voice of the "Stormcock" or Missel-thrush, is now at its best. Cheerily sounds his splendid song, from the topmost branches of the trees; early and late the speckled musician pipes away, neither wind nor storm staying his notes. It is probable that the Missel-thrush may pair during the late autumn or early winter; hence the richness and abundance of his song, and his noisy, quarrelsome, and gregarious habits at this season. The Starling and the Hedge Sparrow are also autumn songsters, and contribute no mean share to the scanty concert of the woods and fields at this season. Another bird that regains its song directly after the moult is the Skylark. There is something delightfully English about the song of this bird—no other music seems so thoroughly in harmony with our peaceful meadows and breezy uplands. It speaks eloquently of freedom, and is one long musical declamation against restraint. His delightful trills are as well known as they are indescribable,

[marginal notes:]
Wren in song, 22nd September.

Missel-thrush in full song, 1st October.

Hedge Sparrow regains song 26th September.

and though they seem more in place above the buttercups and daisies, primroses and daffodils of spring, they are none the less welcome when carolled forth high up in the fresh autumn air over the browns and yellows and purples of the dying year. The Skylark becomes gregarious in early autumn, and continues to live in flocks right through the winter. These flocks are greatly attached to certain localities, and will frequent a suitable field for months—stubbles, which have been sown down with clover, and coarse, weedy pastures having the preference. *Larks in large flocks, 25th October.*

Throughout the autumn months, birds are constantly shifting their ground—seeking out suitable retreats for the coming winter. The moorlands and the mountains are almost deserted; the birds which have bred on these high grounds during the summer, retire to the coasts, the more sheltered country, and the lower valleys. The Meadow Pipits now leave the moors and visit the fields and manure-heaps of the cultivated districts; the Linnets, the Twites, and the Gray Wagtails forsake the gorse, the heather, and the mountain trout-streams, and frequent the fields and the lower reaches of the rivers. The Curlews, the Plovers, and other water-fowl, have sped to the coasts. The Ring Ousel, after staying as long as the mountain fruits lasted, has left the more northern moors, staying a few days on the smaller southern heaths, such as Dartmoor, but all uneasy to be off back to its winter quarters. Many birds *Meadow Pipits return to lowlands, 15th September.* *Ring Ousels leave, 15th October.*

of this species are seen in various wild corners of our southern counties, both in spring and autumn, and a few stray pairs breed in Devonshire and on the Welsh mountains; but the great summer haunt of the Ring Ousel does not reach south of Derbyshire. The Cuckoo has now left the moorlands, and after loafing about the lowlands for a day or so, sets off to Africa. The Merlins have retired to the lower grounds; the Wheatears and Whinchats have commenced their long southern journey, the former birds gradually gathering into enormous flocks as they reach the Downs; and the Red Grouse, the Ptarmigan and the Eagle are almost the only birds left upon the highest uplands. Autumn is the season for many birds to gather into flocks for the winter; solitary birds become gregarious, and most species now display an amount of sociability they never indulge in at any other season. These large flocks are composed principally of the young birds hatched the previous season, and their parents. Another marked and important feature of bird life in autumn, is the gathering together of many of our migratory birds, previous to taking their departure. In the early September days, we see the Swallows and Martins congregating in vast companies at certain well recognised points of meeting. One of these places of which I have a vivid remembrance, was an old flour mill, at the back of which was a large dam whence the supply of water was obtained. Every autumn the Swallows and Martins congregated here in thousands upon thousands, perching on the mill roof, and circling in a countless throng above the pool. High buildings appear to be a great attraction to these birds when gathering together just previous to the autumn flight. I have known lofty factories, churches, and cathedrals visited year by year for this purpose. These Swallow meetings are intensely interesting. The air is full of the fluttering little birds, as they course to and fro, noisily twittering to each other, as if discussing the long journey before them. Rows of tired birds perch on the roofs of buildings, on fences, and the dead branches of trees, on the tops or "rigs" of stacks, and on telegraph wires. Many of these are young birds, and their parents may frequently be seen conveying food to them. For the last ten years I have thought that the Swallow has been gradually decreasing in numbers in certain districts where I formerly remarked its abundance. On the other hand, the House Martin is certainly increasing, and also extending its range. This matter is one for the naturalist to further investigate and observe.

Swallows congregating in large flocks, 1st to 18th September.

As soon as the grain has been cut and harvested we find the stubbles become the chosen haunts of many kinds of birds. One of the most familiar birds of the stubble is the Partridge, now in coveys of varying size. We may often flush these skulking, timid birds from our very feet, and the whirr of their wings is quite startling as they hurry away. Partridges in autumn are very fond of "jugging" in the grass-fields, repairing to the stubbles and turnips to feed. This pretty little game bird very often sleeps upon the stubbles, the covey forming into a circle each bird with head pointing outwards, so that danger is more readily detected from whichever direction it may approach. The rings of droppings unerringly proclaim these roosting-places. Large flocks of Sparrows and Greenfinches also frequent the stubbles to feed on the scattered grain; and where the fields have been sown down with clover, we are sure to meet with the Skylark in abundance. The Yellow Bunting may often be seen on the hedges, especially near the gateways and in the lanes where the bushes on either side almost meet overhead. Here great numbers of straws become lodged, brushed from the waggons as the corn is carried through them; and these little birds are busy picking out the grain, sometimes fluttering for a moment before a full ear then bearing it to the ground. The Wood Pigeons and Stock Doves, Wild Geese and Ducks also come to the stubbles for their share of the corn. I often linger towards evening and watch the wild Pigeons seek their roosting-place. At this season they love to sleep in the fir plantations, coming from the corn-fields in twos and threes and little parties, and settling among the dark green branches with a noisy rattle of their wings. Our resident Wood Pigeons are largely increased in numbers in the autumn by migratory birds from the Continent, great

Flights of Wood Pigeons arrive, 25th October.

AMONG THE BIRDS IN AUTUMN.

"rushes" often making their appearance in districts where food chances to be plentiful.

The turnip-fields are also replete with bird life now. In early morning, as we wander knee-deep among the broad, green leaves, all wet with dew, we flush great numbers of Song Thrushes and "Stormcocks," which visit these places for the slugs and worms; whilst nearer the hedges the Blackbird rises up with noisy clamour, and the Hedge Sparrow flits off to the nearest cover. Where the turnips have partly failed and a bare patch of ground occurs, especially if it be at all swampy, we are sure to meet with the Meadow Pipit, which flits about and perches on the leaves, uttering its plaintive note; and in places where the crop is thickest we may by chance flush a Short-eared Owl, a stray Woodcock, or a Landrail which has not yet left its summer haunts. The Short-eared Owl and the Woodcock both come to this country in great numbers every autumn, arriving on our coasts together. They migrate at night, sometimes in scattered parties or pairs, sometimes in great rushes, according to the state of the weather. The Owls when they reach land generally make for the nearest turnip-fields, hiding themselves under the broad, green leaves; but the Woodcocks, most solitary of birds, scatter the moment they reach land, and shelter for a day in the tangled hedge bottoms among the crumpled russet leaves, or nestle in the fringe of long, coarse grass behind the beach.

Landrails leave, 15th September.

Short-eared Owls and Woodcocks arrive, 25th October to 15th November.

The woods in autumn contain much of interest. When October paints the birch coppices in brightest tints of yellow, the charming little Goldcrest is busy amongst them. Parties of these delicate birds (the smallest British species) linger in them a few days on their southern journey, and every now and then the males burst out into sweetest song. They freely fraternise with the Cole Tit and Blue Tit, but frequent the slender twigs where the seed-pods hang rather than the thick branches and trunks which the Tits search for insect food. Wonderfully tame and confiding are these pilgrim Goldcrests, allowing the observer to watch their every movement as they flit about the trees and bushes. Many of these flocks are bound for more southern regions, and only stay in those birch woods on their line of flight for a week or so; other flocks break up into parties, and disperse over the surrounding country. In the oak woods, where the acorns chance to be the thickest, we shall be sure to meet with shy Pheasants wandering beneath the trees in quest of them; whilst in the branches overhead the Jays and Rooks are pulling them off the twigs. The Rook may often be seen clinging to a large acorn at the extremity of a slender twig, breaking it off with his own weight. Our rambles through the woods will reveal the departure of the summer migrants. No longer does the little Gray Flycatcher moodily sit on the long branches that droop over the wall into the pasture fields, where we saw him so constantly

Migratory Goldcrests arrive, 15th October.

Pheasants over moult 20th November.

Gray Flycatcher departs, 25th September.

in summer; he has sped away to the oases in Algeria. No longer do we hear the trembling notes of the Wood Wren in the tree-tops; the memories of Northern Africa are once more revived in the brain of this little species, and its impulse to return is stronger than the desire to tarry here. The Chiffchaff has sped southwards, although he is the latest of all the Willow Wrens to leave us, as he is the first to come in spring. His monotonous cry is almost hushed, and he takes his departure in silence. The Nightingale has left his haunts in the spinneys and the woodlands by the latter end of September, and the Redstart and the Turtle Dove set off a week before. The fields are losing fast their summer visitors. The Tree Pipits no longer skulk close among the herbage; their moult is completed, and, clothed in their new dress, they are well fitted for the long journey south. The Whinchats, now for the most part solitary, may sometimes be seen among the turnips, or even in the hedgerows. Both these birds are remarkably silent now. In the shrubberies and plantations changes are taking place among the birds. The Chaffinches, after leaving these places all the summer, now begin to flock to them at nightfall to roost. Another bird that regularly makes its appearance in these situations during the autumn is the Redwing, arriving in flocks, and roosting year after year in one particular place. The Brambling is another of our winter visitors from Scandinavian forests, but one that is very irregular,

Wood Wren leaves, 20th September.

Chiffchaff leaves, 4th October.

Tree Pipit and Whinchat leave, 29th September.

Chaffinches over moult, 20th September.

Redwings arrive, 11th October.

Bramblings arrive, 4th November.

both in the time of its appearance and in the districts it frequents, being greatly influenced in the latter habit by the abundance or rarity of beech "mast." Both Bramblings and Redwings frequent the fields in the daytime, especially those where manure is being spread, and seek the evergreens at dusk. An interesting habit of the Rook in autumn must not be overlooked. The naturalist will find that the birds belonging to most if not all of the smaller rookeries desert them shortly after the young can fly, and often join the members of the larger colonies, never going near them again until the autumn has commenced. Then they visit them day by day, generally arriving each successive morning, almost to the minute, at one particular time.

<small>Rooks revisit old nests, 5th September.</small>

Some of the most familiar birds of the commons and the open heaths quit them in autumn for more sheltered districts, or retire across the sea to warmer lands. One of the most familiar of the latter class of birds is the Nightjar, which loses its summer song just previous to moulting, and as soon as that period is passed this species begins its migration to Africa. One of the earlier birds to leave in autumn is the Swift. These curious and interesting birds are particularly active during the calm, warm evenings of July and August, sweeping round the lofty towers and circling high in air above the pools and meadows, uttering their shrill whistling cries as they toy with and chase each other to and fro. The young birds

<small>Nightjar leaves, 22nd September.</small>

<small>Swifts depart, 7th September.</small>

generally leave us a week or more before their parents.

Changes rapid and continuous are also taking place among the birds of the coast. At all the summer stations of the sea-birds, as soon as the young can fly, a grand breaking-up takes place. The long winter vacation has commenced, and the birds of these rocks, and islets, and cliffs disperse over the surrounding sea, many species, such as the Terns, going southwards to the tropics. The low, flat mud-banks and salt marshes of the eastern counties, which in summer were almost deserted by birds, now become thronged with feathered visitors from all parts of the north. One of the first birds to make its appearance on these mud-flats is the Knot. At the close of summer the young birds arrived ; now in the early autumn days the old birds, having moulted into their gray winter plumage ere leaving the high north where they have bred, arrive in thousands. Many of them linger on these muds all the winter, but the greater number work slowly southwards to the west coast of Africa. Dunlins in countless hordes rest on these muds, and may often be seen in compact flocks, wheeling and turning with marvellous precision just above the shallow water or the wet shining beach. Another bird which comes a little later on in autumn to these places is the Hooded Crow. This bird is a visitor from Northern Europe, and picks up a rich living on the coasts and adjoining meadows, often following the course

Terns moving South, 1st September.

Old Knots arrive during September and October.

Hooded Crow arrives, 12th October.

of tidal rivers to more inland districts. The shores of the Wash in autumn are certainly one of the most interesting haunts of birds in this country. Miles of mud-flat and salt-marsh, an endless labyrinth of pools and streams, form a safe retreat for almost every species of British shore-bird, from the big lumbering Gulls and Geese to the shy Ducks, wild Curlews, and dainty Dunlins. The arrival of these birds depends a good deal on the state of the season. If a mild and open one, November is well advanced before the coast is in any way crowded with birds; but in severe seasons they swarm there in October. The fishermen profit by such an abundance of birds, and thousands are netted or shot every year in this one district alone. Miles of netting are stretched across the muds in most parts of the Lincolnshire Wash, and with a suitable tide and favourable weather vast quantities of birds are caught. Many a storm-driven feathered wanderer is observed in this district during the autumn. After severe gales, the various species of Petrel are often taken in the flight nets, and may sometimes be seen flying about in a bewildered sort of a way. Here the Fulmar is often seen, driven in from the German Ocean by the wind, and hopelessly lost in the land-locked Wash. Here the Stormy Petrel and its congener, the Fork-tailed Petrel, lose their way in the darkness of the night and the blinding tempest, sometimes even flying over the land in vain attempt to find the open sea again. Now and then a Northern

Petrels wander south, about 25th October.

Diver, a Great Crested Grebe, or a Little Gull, all on their way south, lose their way in the Wash; and sometimes a Gray Shrike appears, stays a day or so, and then departs. Vast flocks of feathered pilgrims, all bent on getting to their winter quarters, pass this interesting district during the autumn. For days and days together the air will be full of Skylarks; then flock after flock of Golden Plovers cross over, and string after string of Geese and Ducks. Short-eared Owls fly over in the night from the Continent; and Redwings and Fieldfares hurry south with every favourable wind, the former birds every now and then uttering their liquid cry, and the latter *sak sak*-ing to each other as they go. Fieldfares arrive, 28th October.

From this paradise of wading birds, it is but a short distance to the district of the Broads; and here we meet with many birds in autumn that prefer an inland to a littoral haunt. Some of the most charming pictures of bird life may be here witnessed by the observer who is careful not to alarm his feathered favourites. In the evening the big white banks of autumnal fog hang low over the cold, gray waters, and the sighing of the wind through the dead, bending reeds, is fitting music to such a wild, even solemn scene. The shy Wild Ducks are swimming about in the half light, all unconscious of danger; here a brace of Teal float side by side, there a Pochard paddles lazily to and fro; whilst a big, gray Heron, in moody contemplation of all things piscatorial, stands like a statue on an old wrecked fen-boat, half submerged in the stained and stagnant water. Timid Rails and more impudent and trustful Moorhens swim in and out among the reeds and flags; and on rare occasions a Bittern or a Wild Swan will fly startled away. In the bright autumn mornings when the sunbeams play across the water and light up the beds of reeds and rushes, the Broads are made merry and gay with bird life. Little parties of the rare and beautiful Bearded Titmouse cling to the rustling reed stems, and flutter across the open water from one thicket to another, trooping along in a merry, ever-active train; scores of Coots swim gracefully about the wide, expansive pools, holding high carnival in the rushes; and various species of fresh-water Ducks, ever wary and watchful, detect the slightest danger from afar.

Throughout the autumn months the migration of birds is in progress. The spring movements of birds are much quicker than in autumn. Then birds seem all anxious to get to their summer quarters as quickly as possible; but on the return journey they travel much more leisurely, and, of course, in larger numbers. Birds are also more gregarious in autumn than in spring, and many species fraternise for the journey which are never seen in company at any other time. Night and day these little feathered travellers are hastening southward—following summer in one vast fluttering throng. In the still autumn nights we may often hear the cries of these migratory birds, as they pass over high in air above us. Sometimes the harsh notes of a Heron will be heard, or the shrill call of a Curlew or Godwit; and now and then the loud trumpet-like scream of the Wild Swan, rendered musical by distance, sounds clearly from the sky.

The migration of birds to and from this country is of several kinds. Many birds come to us in autumn, and remain throughout the winter, such as Redwings, Fieldfares, Bramblings, Woodcocks, and Jack Snipes. Others merely pass along our coasts on their way still further south, or cross the inland districts on their journey. The pretty Dotterel, having safely reared its brood on the very summits of our northern mountains, now speeds back to Africa. This bird especially seems to travel slower in autumn than in spring; at the latter season it may even perform the astounding feat of travelling from Africa to England in a single night! How utterly insignificant do our lines of fast mail steamers, even our flying railway trains, appear in comparison with such speed as this! The Dotterel is a most tame and confiding little creature, and will allow you to watch its movements without the least show of shyness or fear. The Stone Curlew now quits the heaths and wolds where it rears its young, slipping quietly and stealthily away on some bright, moonlight night, when all is favourable for its southern flight. Another bird that passes this country very regularly in autumn is

Jack Snipes arrive, 15th October.

Dotterels leave, 29th September.

Stone Curlew departs, 15th October.

the Rough-legged Buzzard. This species also has its regular lines of flight, and travels along them, scarcely swerving half a mile from its course year after year. It follows the mountains, passing right down the backbone of Great Britain to the sea, whence it crosses to its winter haunts in the south of Europe. Many of the smaller birds follow an inland course until they reach the coast. Birds in autumn may suddenly become numerous in certain woods and fields, remain a few days, then as quickly disappear. This is especially noticeable with the Song Thrush, the Goldcrest, the Chaffinch, and the Wheatear. Some of these flocks of Chaffinches are very large, and what is rather curious is that the earliest companies are almost invariably males, the females arriving a week or so later. The other kind of migration is the usual departure of those birds which visit us in spring, and live with us throughout the summer. The earliest birds to come, such as the Willow Wren, the Blackcap, the Wheatear, and the Sand Martin, are the latest to depart; whilst those that arrive on our shores when spring is well advanced —such as the Cuckoo, the Swift, the Gray Flycatcher, and the Reed Warbler, are the first to hurry away at the close of summer.

The migration of birds is beset with many perils and many difficulties. Birds often lose their way; a contrary wind or a spell of dark, cloudy weather appears to disorganise their movements, and, like mariners without a compass, they are at a loss which direction to take. Many wonderful scenes are witnessed at the lighthouses on some parts of the British coasts during the season of migration in autumn. Sometimes when the moon is suddenly hidden by a bank of clouds, the lanterns of the lighthouses are the points to which the stream of migrants hasten, and where, in a confused fluttering throng, they beat against the glass like moths round a street lamp, and fly to and fro, utterly bewildered and completely lost. They seem to have no idea of their true course, and fly aimlessly about, many killing themselves against the glass, others falling into the water below. The light-men are alert on these occasions, and capture numbers of the poor, lost travellers. Many of the birds are too tired or too dazed to move, and allow themselves to be taken by the hand as they sit on the balcony. If the scene is such a vivid, interesting one, in the small space round the lantern, how much more so would it be could we only see the vast feathered army in full progress across the night sky, when the tide of migration is at its height! Here, where the lanterns are the central point of attraction for the countless hosts when the sky becomes overcast, the sight is wonderful in the extreme. Birds of many different species are flying together, or are attracted from all points of the compass by the brilliant light. Ducks and Geese are travelling with Goldcrests and Swallows. Starlings and

Finches are flying side by side with Gulls and Waders. Warblers and Herons scatter scientific classification to the winds, and fraternise with Swans, and Goatsuckers, and Larks. Falcons and Owls appear to lose all propensity for preying on their helpless fellow voyagers, and fly harmlessly to and fro amongst their companions in misfortune. The light is literally vignetted in drifting masses of birds, which appear an instant as they cross the brilliant rays, and then suddenly vanish in the gloom; but as soon as the weather clears and the clouds break, and the moon shines forth once more, the birds appear to get on their right track again, and the feathered hosts are gone as suddenly as they came. I always think that these migratory movements lend bird life its greatest charm in autumn. Whenever and wherever we wander out into the woods or by the shore, signs of the great feathered exodus now in progress are to be seen on every side. In the highest air the V-shaped lines taken by the migrating Geese, or in the lower atmosphere the fluttering throng of Martins and Swallows, proclaim the movements of some of our feathered friends. Almost every wood, and field, and lane, and highway are tenanted by birds bent on their annual pilgrimage; and bird after bird departs as the autumn draws to its close, their places being taken by others from more northern lands.

The habits of the Starling right through the autumn, very closely resemble those of the Rook at this season. This bird lives in flocks like Rooks, some of its gatherings being very large. These companies are remarkably regular in their movements, having certain recognised places where they assemble at morn and even; whilst during the day many of the old birds pay visits to their old breeding-places, the males singing lustily from time to time. Now the young males may be heard making attempts at song. Some of these even in the middle of November have not fully acquired the brilliant dress of maturity, the plumage on the head being changed last of all. These young birds sing in a very low strain, and their music is nothing near so varied as that of the adults—another proof, by the way, that the songs of birds have to be slowly learnt in infancy, just like the language of mankind. Then, again, there is something ineffably sweet about the Robin's song at this season, especially as we listen to it at nightfall, when the evening mists are rising from the woodlands, and the setting sun reflects a refulgent light over the brightly-painted trees. It is a song of hope, among the ruins of summer's brilliant pageant, and a prophecy that, beyond the coming winter, spring with all its glories will again return.

Young Starlings singing, 25th September.

Then down by the stream side in the alder trees the Siskins and the Titmice are busy. All the summer the former little birds have been in the northern fir forests, now they seek the lower and more sheltered districts. Again, how engaging are the Redpoles and Goldfinches on the wild, weedy wastes and down the hedgerows. In a trooping train the Goldfinches flit from stem to stem of the tall, prickly thistles, scattering the downy seeds to the winds as they go; and the Redpoles, in a compact little flock, drop down from the hedges or the tree-tops, to pick out the tiny seeds among the coarse herbage. In the autumn, numbers of young Herons wander far from their birth-place and stray up the rivers and brooks in quest of fish. They are nothing near so wary as their parents, and but very few survive all the perils of their youth. The Gray Wagtails have deserted the northern trout streams; and from the shores of the upland pools, the common Sandpiper has now taken its departure. Numbers of these birds, however, appear to pass over our islands during the month of October from more northern haunts.

Common Sandpiper leaves, 15th September.

Everywhere we turn all Nature, and more especially bird life, seems at rest after the bustle and excitement of spring and summer; a dreamy quietness prevails, and the few songsters that warble now only seem to emphasize the tranquillity of the autumn woods. Nut-brown October is one of the pleasantest months in the whole year for the naturalist, and from the beginning to the end of this delightful period bird life is for ever on the change. As the painted leaves fall to the earth and accumulate in big brown and yellow drifts by the tree-trunks and in the hollows, we are enabled to observe the actions of many birds that all the summer have been hidden by the foliage. Every day the branches become more bare, and show out naked among the sprinkling of leaves. We can now watch the Woodpeckers pursue their erratic way up the gnarled and rugged trunks in quest of insects, and scan every movement of the Tits and Ring Doves high up in the slender branches. Then on the ground how the various birds betray their whereabouts as they hop among the dead, fallen leaves. The Thrushes and the Blackbirds particularly love to frequent the ground at this season, and hop along under the dark shade of evergreens and hazel bushes with surprising quickness. In our wanderings through the plantations and quiet corners of the coppices we may now flush the Woodcock from its retreat amongst the dead leaves. The Pheasants rise on whirring wing and top the brushwood as they hasten from the fields to the covers. Ever timid and fearful, these handsome birds dislike to be surprised in the open, and will often run with great speed through the herbage and enter the covers through a rabbit run or even under the arch of the stream. Wherever ploughing is going on in the late autumn we may be pretty certain of meeting a variety of birds. Missel-thrushes love to search the newly-turned ground for worms and grubs; parties of Wagtails catch the insects; and

Rooks wander up and down in quest of food. In some districts, especially in the low-lying eastern counties, the Black-headed Gull is a constant companion of the ploughman, and searches among the soil for similar substances to those eaten by the Rook. It is a charming sight when a large flock of these beautiful birds rise in a fluttering throng and fly off to a distant part of the field, or even settle on the tops of the trees in the hedges. Black-headed Gulls wander far and wide from their breeding stations as soon as the duties of the year are over, not only visiting the coasts, especially those which are low and muddy, but following the course inland of tidal rivers for miles, and spreading out over all the surrounding farms. Lapwings also congregate on the newly-ploughed land, and are especially fond of frequenting flooded meadows, where an abundance of food is obtainable. These birds leave the moors in autumn and become even more gregarious during this season and the winter. A considerable migration of Kingfishers may be observed during October, especially in the low-lying counties near the Wash and along our southern coasts. Most of these migrants are young birds, easily recognised by the dark brown tinge on the breast.

As in spring and summer, the following table will show the state of bird music during the autumn months :

AMONG THE BIRDS IN AUTUMN. 191

NAME OF SPECIES.	SEPT., 30 days. SANG ON	OCT., 31 days. SANG ON	NOV., 30 days. SANG ON
1. Missel-thrush	30 days	31 days	30 days
2. Song Thrush	Silent	Now and then	Silent
3. Blackbird	Silent	Very rarely	Silent
4. Robin	30 days	31 days	30 days
5. Wren	Last 8 days	31 days irregularly	30 days
6. Hedge Sparrow	Last 4 days	31 days irregularly	Irregularly according to weather
7. Willow Warbler	Last 5 days	First 2 days exceptionally	Absent
8. Chiffchaff	Last 15 days	First 3 days	Absent
9. Whitethroat	Silent	Absent	Absent
10. Blackcap	Silent	Absent	Absent
11. Redstart	Silent	Absent	Absent
12. Starling	Occasionally	Occasionally	Occasionally
13. Meadow Pipit	Silent	Silent	Silent
14. Tree Pipit	Silent	Absent	Absent
15. Chaffinch	Now and then during last 10 days	Very exceptional	Silent
16. Yellow Bunting	30 days irregularly	Now and then	Silent
17. Greenfinch	Last 15 days occasionally	Not unfrequently	Silent
18. Skylark	Occasionally	Occasionally	Occasionally
19. Cuckoo	Silent	Absent	Absent

From these data it will be seen that the Missel-thrush, the Robin, the Wren, the Starling, and the Skylark, are the most prominent musicians during the autumn. The Hedge Sparrow is very irregular in his song, being influenced by the weather and the nature of the haunts he frequents. If these are at all bare and exposed he sings little; and even in the shrubberies a week or more often passes without him engaging in song. In October the Song Thrush warbles occasionally; and the Willow Wren and Chiffchaff, as soon as they get over the moult, sing from time to time, until they leave for Africa. The Chaffinch breaks out into fitful song after the moult; and the Yellow Bunting and the Greenfinch also help to swell the meagre concert of the woods and fields at this season.

Our wanderings among the birds in autumn are full of uncertainty, which tends to increase the charm of outdoor observations. Now is the season for some of our rarest birds to accidentally reach this country—distinguished strangers from distant lands, who have lost their way, and wandered from their usual course. We have no room here to notice them, but will reserve the following chapter for that purpose. Many of the birds that lent an interest to the woods and fields in spring and summer have gone; but, to compensate us for their absence, others have come, and are constantly arriving, to take their place; whilst this element of uncertainty as to what species we may chance to meet with in our rambles is a grand inducement to spend as much as possible of the delicious, dreamy autumn out of doors among the birds.

CHAPTER III.

STRANGERS OF THE AUTUMN.

THE number of birds that are regarded by naturalists as British amounts to nearly four hundred species. The claims of many of these are very slender, either resting on doubtful authentication, or only on one or two accidental appearances. Some there are that have undoubtedly escaped from captivity; others appear more or less regularly during the periods of migration, more especially in autumn; whilst a few only visit us at long and uncertain intervals. Of these four hundred species nearly half are only accidental visitors, the bulk of them from the Continent of Europe, but a considerable number from North America, and the least of all from Asia. All these feathered strangers are migratory, and are individuals—generally young ones—that lose their way in travelling to and from their winter home. The great highways along which migrants travel are extremely complicated, and, to understand the occasional visits of some of these birds, we must make ourselves acquainted with the paths by which they come. The general movements of migratory birds are very different in spring to what they are in autumn. In spring, as we have already seen, the great migration is from south to north. No bird is known to migrate north in autumn, but at that season the great movements are from north to south, from north-west to south-east, and from north-east to south-west. Some of our autumn strangers reach the British coasts during their usual passage south; others, that should travel from the extreme west to the far east, occasionally take the wrong direction, and stray down the Atlantic coasts of Europe; whilst a great many European species at this season travel from the east to the west; and sometimes an Asiatic species gets mixed up with them, and is borne westwards with the stream of regular migrants. It is most probable that only the very outermost edge of this vast wave of migrants breaks upon the extreme west of Europe and the British Islands; so that, though there may be a great number of Asiatic birds carried with it, the bulk of them are scattered over Central and Southern Europe. Most of these birds appear to cross the German Ocean by way of Heligoland. The occurrence of a rare bird on this famous little island at the mouth of the Elbe is a great point in the probabilities of another of that particular species turning up in this country, and helps materially to confirm the *bona fides* of any stray individual whose occurrence here may be attended with doubt. When we know that these distinguished strangers occur from time to time on Heligoland, we consider it very probable that others of that particular species may come across to us with the regular stream of migrants; for there is an enormous passage of birds over this favoured little spot to the British Islands every autumn. Any bird that may chance to wander to Heligoland may therefore just as likely pay us a visit too. On the other side of the Atlantic, the Bermudas and the Azores may be aptly termed the American Heligoland; and practically all the New World birds which have visited the British Islands, the genuineness of whose occurrence is not open to doubt, are more or less regular birds of passage over them in autumn or spring.

In order to understand the philosophy of this bird errantry still farther, it is well to point out what a great number of strictly West Palæarctic birds reach the limit of their eastern range in the valleys of the Yenesay, and possibly the Lena, in Central Siberia. Most of the birds that travel so far east in spring, return to South Europe and Africa to winter; consequently, there is a very important stream of migration in this direction in autumn, which drains off a few of the strictly Asiatic species whose proper destination at that season is Turkestan, Persia, India, China, the Burmese Peninsula, and Australia. During the last glacial epoch, the entire bird population of the northern parts of the Eastern hemisphere was driven southwards into Africa, India, China, Burma, the Malay Archipelago, and Australasia. These countries to this day are the grand winter home of all the migratory birds of the northern portions of the Old World; the European species crowding into Africa, the Siberian ones into India and the other countries in the south. Some of these European birds have increased so much that they have been compelled to spread eastwards into Asia; whilst some of the Siberian species have extended their range westwards into Europe from a similar cause. It is therefore easy to understand how some of these individuals or their offspring, which have increased their range westwards from Asia, wander down the coasts of Europe to England; whilst the individuals of the European species that go to Asia in summer, entice a few of the indigenous birds back with them to Europe and Africa. Having thus made ourselves familiar with the principal movements of birds at migration time, we will proceed to notice those species which from time to time appear upon our coasts as "strangers of the autumn."

Perhaps by far the most interesting of our casual visitors have made their appearance in the autumn. We will take them in the order of their usual classification, beginning with the Birds of Prey. The occurrences of the Griffon Vulture, and the Egyptian Vulture, were all in autumn; and doubtless these birds were blown north by gales either from the Canary Islands or the Spanish Peninsula. It must also be remembered that a flight up the English Channel is little more than a morning stroll for a Vulture. The Jer Falcons are thoroughly Arctic birds, and occasionally wander southwards in the autumn. These are mostly birds of the year, either roaming about in quest of suitable haunts, or tempted to follow the vast flocks of Arctic birds that draw south at this season. The Spotted Eagle occasionally visits us at this season, from its haunts in the vast forests of Germany. Our next noteworthy species, is the Red-footed Falcon. This bird spends the summer in the forests of Hungary, Russia, and South-Eastern Siberia, and retires in winter to Africa. Though it is said to pass through Germany on migration, it is more than probable that the individuals which visit the British Islands, are merely stragglers in the great stream of migrants from the east. Careful attention should be paid to any examples of this bird that may chance to be captured in this country, as it is just possible they may be the eastern form of the Red-footed Falcon, which breeds in Eastern Asia, and winters in India and South Africa. Another bird to be looked for in autumn, is the Lesser Kestrel. Although this species has only once been obtained in this country, it is very probable that it has often been overlooked, or confused with the Common Kestrel. Like the preceding bird, this species also appears to be drifted westwards in autumn.

The Spotted Eagle, the Goshawk, the American Goshawk, and the beautiful Swallow-tailed Kite, are all autumn wanderers to this country, the two former from the Pomeranian forests, the two latter from North America. Three of our rarest Owls have visited us at this season likewise: the Snowy Owl, the Hawk Owl, and Tengmalm's Owl, all wanderers from the Arctic forests and tundras.

We now pass on to that important family the Passeridæ, amongst which we shall find many instances of bird errantry. The birds in this family are famous for the great journeys they perform in spring and autumn, the majority of the species being more or less migratory. First to claim our notice is White's Thrush—a bird all the way from Eastern Siberia and China! Most of the examples obtained in this country were got in winter; but unquestionably they came with the autumn flight of birds from the east, and had loitered on the way. Another Siberian stranger which may be looked for any autumn, is the Black-throated Ousel. The single example of this species hitherto obtained, like the preceding, also lingered on the way across Asia and Europe, and did not reach this country apparently until the winter. Another autumn wanderer to the eastern coasts of the British Islands, is the Scandinavian form or sub-species of the Dipper. It may turn up any season with the migrants from the north of Europe, and should be carefully looked for in all suitable localities. Another little Arctic stranger which appears from time to time in this country in autumn, is the Blue-throated Warbler. It sometimes appears in considerable numbers, and as the females and birds of the year are dull-plumaged little creatures, they may often be overlooked. This bird usually occurs in September and October. Three species of Chats have also visited us at this season: the Desert Wheatear, the Black-throated Chat, and the Isabelline Chat. These birds most probably came to Western Europe with the crowds of migrants that pass west from Siberia and Turkestan. The Red-breasted Flycatcher is a stranger of the autumn. This bird breeds in the German forests, but is rare in the extreme west of Europe, so that I am of opinion that the stray individuals which have reached us came from the east. The occurrence of several examples of the Rufous Warbler in this country in autumn, is an interesting, yet puzzling circumstance. This bird only breeds in the south and east of Europe, and in Palestine, therefore it has no inducement or reason to go north of these localities in autumn. There would be nothing extraordinary in an example of this species straying to the British Islands in spring, as birds of many kinds are apt to overshoot the mark at that season, and get far north of their proper habitat. Nor would it be difficult to comprehend how the eastern race of this bird, which is found between Greece and

Turkestan, might wander here in autumn with the usual stream of migrants bound west; but the examples hitherto obtained are all said to belong to the western form, although the eastern race has been met with at Heligoland. A good look-out should be kept for this bird in autumn, especially on the south coast of England.

One of the most interesting birds that has ever visited this country from the far east is the Yellow-browed Willow Wren. This delicate little bird lives in the pine forests of Eastern Siberia, and winters in China, Burma, and India; yet every autumn a considerable number appear to lose their way, and, instead of going direct south to their winter quarters, join the western stream of birds, and after travelling with them nearly four thousand miles, pass Heligoland, apparently on a southern course towards Africa. It would appear that some of these lost little birds survive the winter, and actually go back in spring—or attempt to do so—by the same road that they came in autumn! Another tiny autumn stranger is the Firecrest; and at the same season observers should keep a sharp look-out for stray examples of the Continental or Arctic form of the Long-tailed Titmouse. Now and then the Alpine Accentor visits the British Islands. Probably the examples which reach us do not come from any part of Europe (throughout which, in the districts it frequents, it is a resident, only descending the mountains in winter, like our own Meadow Pipit), but from

Turkestan. Nutcrackers, though for the most part resident, occasionally visit the British Islands from the forests of Northern Europe; but they, like the beautiful Waxwing, are only gipsy migrants, and appear at very uncertain intervals. Pallas's Gray Shrike and the Great Gray Shrike are also accidental visitors to this country in autumn, the former from the forests of Siberia, the latter from Northern Europe. Gray Shrikes obtained in the British Islands require careful identification, for these two species are very easily confused.

Specially interesting are the autumn movements of the Rose-coloured Pastor. This remarkably handsome bird appears accidentally in the British Islands—usually young ones which, having had no experience of the road, lose their way and wander aimlessly about. The Rose-coloured Pastor is one of those birds whose migrations are very exceptional, journeying eastwards in autumn to its winter home in India. This species well illustrates the complexity of the various movements of birds; and on its way east to India must absolutely pass many other species which are travelling just as directly west! Other gipsy migrants which turn up from time to time in the British Islands are the Pine Grosbeak and the White-winged Crossbill from the northern forests of Europe. The American form of White-winged Crossbill also visits this country by way of Greenland, and doubtless it has been much confused with its European representative. One or two examples of the Scarlet Rose Finch have also occurred in this country. It is a most interesting little bird, and breeds as close to us as the Baltic Provinces, yet goes all the way to India to winter! Odd birds now and then take the wrong direction, and wander west instead of east. One other Finch should be looked for in this country, as most likely to appear in autumn, and that is the eastern form of the Common Bullfinch, to which naturalists have assigned the name of *Pyrrhula major*. Stray examples of this race repeatedly wander to Germany; and there is nothing to prevent them being brought by the western wave of migration from the east to the British Islands.

The researches of the late Mr. Swaysland, of Brighton — whose accomplishments as a field naturalist have been carefully and wilfully ignored by his "scientific" but far less learned contemporaries, who have profited largely by his knowledge—have shown that the Wild Canary and the Serin are accidental wanderers here in autumn. The former species is probably blown north from its island home by exceptionally severe gales; the latter is an accidental straggler from the woods of Western Europe, probably by way of Heligoland. Among the various Finches that visit us in the autumn must be mentioned the Mealy Redpole, an inhabitant of the Arctic forests; its close ally, the Greenland Redpole, has also been obtained in this country at the same time of year. Three other strictly eastern species of birds, which have wandered as far west as the

British Islands, are the Lapland Bunting, the Little Bunting, and the Rustic Bunting. The first-named species breeds on the tundras of both hemispheres, beyond the limits of forest growth, the eastern hemisphere birds wintering in Mongolia and China; so that individuals of this species that breed as far west in Europe as Norway, cross nearly four thousand miles of country to reach their winter quarters! The two latter species are found breeding in high latitudes, from Archangel to the Pacific; but, like the Lapland Bunting, even the European individuals cross the two continents and winter in China! All these Buntings occasionally miss their way in autumn, and wander south into Central Europe, and accidentally stray—perhaps by way of Heligoland—to the British coasts. The Ortolan Bunting is another species of this group which strays here in autumn—most likely from Scandinavia, when on their way to their winter quarters in West Africa. The Black-headed Bunting has visited us from Southern Europe.

We now pass on to the Pipits and the Larks, amongst which, some of the most interesting strangers of the autumn will be found. The first we have to notice is the Red-throated Pipit, a bird that is very probably overlooked here in autumn, as it is well known in Heligoland. It is not a west European species at all, rather a Siberian one, becoming rarer and more local this side of the Urals, and wintering in north-east Africa. Stragglers, however, pass down into West and South Europe, and thus reach this country. Richard's Pipit is a Central Asian species, wintering in India, Burma, and China; but every year numbers take the wrong direction (chiefly young birds), fly west instead of east, and thus get into Europe. The Tawny Pipit visits us occasionally on its way down from its northern haunts to Algeria; whilst the Alpine Pipit has been observed more sparingly. This latter species is not so much of a migrant as the others, generally leaving the high mountains in autumn, and descending to the lower valleys. It is well known on Heligoland at this season, and probably has found its way from there to the British coasts. Five species of Larks are included in our list of rare autumn visitors. The Short-toed Lark, the Calandra Lark, and the Crested Lark occasionally visit us, the two former most likely from Turkestan, as their habitat in Western and Southern Europe does not extend sufficiently far north to place them in the southern stream of migrants; the latter probably comes by way of Heligoland from Scandinavia. The handsome White-winged Lark has, through the acumen of the late Mr. Swaysland, been detected as a visitor to this country. It is an inhabitant of the eastern steppes from Russia to the Altai mountains in Siberia. In the latter country it is a regular migrant, and in journeying to Turkestan to winter, a few stray birds are carried westwards by the great wave of migrants into Europe. There can be little doubt that other examples will occur from time to time, as the bird is occasionally seen in Western Europe and has been taken on Heligoland. The last species with which we have to deal is the Shore Lark. Like the Waxwing and the Snow Bunting, this bird is only a gipsy migrant, having no settled winter home, wandering south with hard weather, and returning north as soon as the storms are past. It comes from the tundras of the Arctic regions, sometimes in considerable flocks, and appears to be of much more frequent occurrence than formerly, though this may be because there are more people on the look-out for rare birds than used to be the case.

Three examples of the Needle-tailed Swift have been obtained in the British Islands. This bird belongs to a little group of species remarkable for the peculiar structure of the tail, the shafts of which are prolonged beyond the web into sharp spines. The Needle-tailed Swift is another bird inhabiting the far east which occasionally takes the wrong direction, and finds its way to Europe instead of Australia in autumn. One of the British examples of this bird is in the possession of my friend, Mr. Tomalin, of Northampton. It was shot upwards of fifteen years ago, in the autumn, at Great Hawksley, in Essex.

The Purple Heron and the Glossy Ibis are only found in Europe during summer, and in autumn occasional stragglers wander as far as the British Islands, where they are most frequently observed in the eastern and southern counties of England. The Purple Herons are mostly in the plumage of immaturity—young birds which wander far and wide as soon as they are able to forage for themselves. Another distinguished stranger from the east is Macqueen's Bustard. The range of this species extends sufficiently far north in Asia to place many of the birds within the influence of the great western stream of migration, which brings these stragglers into Europe. It is also extremely probable that the many examples of the Cream-coloured Courser which have been obtained in our islands are birds borne on this western wave from the far east, although the one or two specimens which were captured in spring may have been birds which have overshot the mark or been blown from the Canary Islands by heavy gales.

It is a curious and interesting fact that most of the accidental visitors belonging to the Charadriidæ, or Plovers and Snipes, which visit this country in autumn belong purely to North American species. The Gray Phalarope and the Asiatic Golden Plover are about the only exceptions; and of these the former is circumpolar, and the latter is represented on the American continent by a closely allied form, which has itself been known to visit the British Islands. The Asiatic Golden Plover is an East Siberian bird which winters in India, Burma, China, Australia, and the islands of the Pacific. It is evident that stray examples wander south-westwards in autumn instead of south. These are probably birds that breed in the valley of the Yenesay, and which get into the stream of migrants travelling west from that region towards Europe. Of these American species we may mention the Killdeer Plover, the Esquimaux Curlew, Bartram's Sandpiper, the Solitary Sandpiper, the Spotted Sandpiper, the Yellow-legged Sandpiper, the Red-breasted Snipe, Bonaparte's Sandpiper, the Pectoral Sandpiper, the American Stint, and the Buff-breasted Sandpiper—all as strangers of the autumn in the British Islands. These birds are for the most part inhabitants of the Arctic tundras of the New World, speeding south in autumn— young and old in vast crowds crossing the seas, and passing along the coasts, as well as down the inland valleys and by way of the Lakes. The Bermudas and the Azores, Greenland and Iceland serve as stepping-stones to Europe. Whilst dealing with American species, it is well here to mention that the Purple Martin, the Belted Kingfisher, the Yellow-billed Cuckoo, and the American Bittern are also accidental wanderers from the New World to the Old in autumn.

Birds of the Gull tribe are also great wanderers in autumn. Such boreal species as Sabine's Gull, Ross's Gull, the Glaucus Gull, the Iceland Gull, and the Ivory Gull are all apt to wander south during the autumn and winter, and stray individuals occasionally visit the British seas and coasts. The Little Gull is a regular summer migrant to the lakes of West Russia; and it is most probably stray birds from these localities that wander to our coast in autumn. The most interesting fact about the distribution of the Little

Gull is that it breeds from West Russia to Northeast Siberia, even to the shores of the Sea of Ochotsk, yet returns to winter in South Europe, Africa, and the extreme south-west of Asia! This shows us how conservative birds are—how strong the impulses must be which attach them to certain districts, in obedience to which they cross and recross a mighty continent instead of seeking a winter refuge in the seas round Japan and China, not one-quarter the distance! Two other species of the Gull family are accidental visitors here in autumn, viz., the Pomarine Skua and Buffon's Skua. Both are circumpolar birds, breeding on the tundras beyond the Arctic circle. The former species is found most frequently on our coasts, in some seasons occurring in considerable flocks. The Little Auk and the White-billed Diver are two more Arctic birds which wander south to the British Islands in autumn and winter. Of the Petrel family, the Sooty Shearwater is the most distinguished visitor in autumn. This species belongs to the southern hemisphere, and wanders northwards, many individuals overshooting their mark and straying to various parts of the North Atlantic coasts, both of the Old World and of the New.

With a notice of the various species of Ducks which have from time to time paid their irregular visits to our islands in the autumn, we will bring the present chapter to a close. Bewick's Swan is a more or less frequent straggler here at that season, occurring most frequently on the wild west coast of Ireland. Its nearest breeding-place is probably Nova Zembla, and it would appear that the birds that summer there go well out to sea, and possibly come by way of Spitzbergen and Iceland, shunning the coasts of Norway, and never appearing to join the stream of migration which flows down the Baltic from the shores of the White Sea. One or perhaps both of the two forms of Snow Goose have been met with in the British Islands during autumn; and as these individuals have been captured in Ireland, it is possible they may have come across the Atlantic. Both the races of this bird may be circumpolar, although, so far as is at present known, the breeding range of the larger form is confined to the Hudson's Bay Territory, and the smaller form to North-west America. The Red-breasted Goose occasionally wanders to our shores in autumn, generally in the company of Brent and Barnacle Geese. The breeding range of this bird is very restricted, probably being confined to the valleys of the Obb and the Yenesay, within a few hundred miles of the coast of the Arctic Ocean. Stray examples of the Ruddy Sheldrake have been met with in this country in autumn. This bird is a resident in those parts of Europe it inhabits, but is a migrant in Asia, so that it is most probable the individuals occurring here have been brought from the far east with the stream of western migrants. The Red-crested Pochard is an accidental wanderer here from Germany; and Steller's Eider, a very rare visitor indeed, from the coasts of Russian

Lapland. The remaining five species are all stragglers from North America, and presumably come by the same routes that the Sandpipers do. They are the American Teal and Wigeon, the Buffel-headed Duck, the Surf Scoter, and the Hooded Merganser.

Scarcely an autumn passes that does not bring some of these distinguished strangers to our shores; and doubtless there are many other species which pay us their uncertain visits, and remain undetected. Almost every year the number of birds noticed in the British Islands receives an addition; it is only fair to presume that this is owing to the greater number of field naturalists abroad nowadays, rather than to any increased migration of these feathered waifs to our shores.

A considerable number of these distinguished visitors come to our islands on both spring and autumn migration; but as a rule most of these are birds that breed north of the British Islands. The Lesser Gray Shrike, the Wall Creeper, and the Roller have visited us at both seasons; as likewise have the Black Stork and the White Stork. The Little Crake and the Pratincole, the Black-necked Grebe, and the Caspian Tern are also common to both seasons; so, too, are Temminck's Stint, the Green, Wood, and Broad-billed Sandpipers, the Dusky Redshank, and some few others. It will thus be noticed that most of the Passerine strangers are either wanderers here in autumn or in spring, few occurring at both seasons of migration. These may be identified as follows:

Species.	Points of Distinction.
Lesser Gray Shrike	Similar in colour to Great Gray Shrike. Forehead black; rosy tinge on breast and flanks; single wing bar. Length 9 inches.
Wall Creeper	General colour bright slate gray; wing coverts, and all except first three quills crimson on basal half of outer web; throat, breast, and tail black, the latter tipped with gray. Female similar to male in colour; but both sexes in winter have the throat white. Length 6 inches.
Roller	General colour of adults verdigris green; mantle and innermost secondaries brown; terminal half of wings black; shoulders and under surface of quills purplish blue. Length 12 inches.
Black Stork	Under parts of adults below neck white, remainder of plumage black, shot with green, purple, blue, and red. Bill, legs, and feet scarlet. Length 40 inches.
White Stork	General colour of adults white, except wings, which are black. Bill, legs, and feet scarlet. Length 42 inches.
Little Crake	Similar to Baillon's Crake in colour, but outer web of first primary brown, instead of white as in that species. Length 7½ inches.
Pratincole	General colour above, brown; rump, upper tail coverts, and basal portion of tail white; axillaries chestnut; tail forked; black line

Species.	Points of Distinction.
Pratincole (*continued*)	beginning at base of bill encircles chin and throat, which are buff; remainder of under parts nearly white. Female duller than male. Immature birds have pale margins to feathers of upper parts; throat streaked; girdle only indicated by spots. Length 10 inches.
Temminck's Stint	Outer tail feathers white. Length 5¾ inches.
Green Sandpiper	Above, brown spotted with white; rump and outer tail feathers white; axillaries and under wing coverts dark brown, narrowly barred with white. Length 9½ inches.
Wood Sandpiper	Similar to preceding, but conspicuous white wing bar; axillaries white, with few faint brown bars. Length 7½ inches.
Broad-billed Sandpiper	Bill flat and wide up to tip. Upper tail coverts and secondaries uniform brown; middle toe and claw ·9 inch. Length 6½ inches.
Dusky Redshank	Distinguished from Redshank and Greenshank by white secondaries barred with brown. Bill black, red at base of lower mandible; legs and feet red. Length 12 inches.
Caspian Tern	Largest British Tern; tail only slightly forked. Length 19 to 21 inches.
Black-necked Grebe	Head and neck black; white on primaries as well as secondaries. Bill recurved. Length 12 inches.

The identification of these rare migrants is often a matter requiring great care and skill, as they are mostly young or immature birds, or, if adults, in winter plumage, when the various points that distinguish them from allied species are not so conspicuous or pronounced. There can be little doubt that many of our familiar European species, whose range in summer extends eastwards into Asia, get lost in their turn, and wander far away into the remote east and south during autumn. Unfortunately there are so few observers in these eastern countries, that such accidental wanderings have every chance of being overlooked. The following table will help the student to identify any of these strangers of the autumn that may chance to fall in his way:

Species.	Points of Distinction.
Griffon Vulture	General colour fulvous brown; wings and tail brownish black; head and neck covered with white down. Length 40 inches.
Egyptian Vulture	General colour white; primaries black, secondaries brown; fore part of head and neck bare of feathers. Length 25 inches.
Red-footed Falcon	Adult male uniform slate gray; thighs, vent, and under tail coverts chestnut. Female, upper parts lighter and barred with grayish black; under parts dul chestnut. Young distinguished by row of oblong white spots on primaries, and chestnut thighs. Length 12 inches.

Species.	Points of Distinction.
White Jer Falcon	General colour white, spotted more or less with dark brown; tail white at all ages. Bill pale yellow. Length 20 to 23 inches.
Iceland Jer Falcon	An intermediate form between the White Jer Falcon and the Brown Jer Falcon, said to have the flanks barred and the bill dark horn colour at all ages. Length 22 inches.
Brown Jer Falcon	Head nearly uniform dark brown. In all three forms the outer and inner toes are nearly equal in length. Length 21 inches.
Lesser Kestrel	Similar to Common Kestrel in colour, but smaller; adult male without spots on back; claws pale yellow. Length 12 inches.
Spotted Eagle	General colour dark brown. Young with yellowish nape patch and white tips to scapulars and wing coverts; under parts streaked with brown. Length, male 24 inches, female 26 inches.
Goshawk	Upper parts grayish brown; tail with four bars of dark brown; under parts white barred with brown. Young brown above, buff below streaked with dark brown; tail bands five in number. Length, male 19 inches, female 24 inches.
American Goshawk	Similar to preceding, but under parts marbled instead of barred with brown.

Species.	Points of Distinction.
Swallow-tailed Kite	General colour white and black shot with blue and purple; tail acutely forked. Length 21 inches.
Snowy Owl	White, more or less spotted, and barred with blackish brown; feet feathered to claws; no operculum. Length 22 to 27 inches.
Hawk Owl	Above, dark brown spotted with white; below, white barred with reddish brown; tail long, barred and tipped with white; feet feathered to claws. Length 14 inches.
Tengmalm's Owl	Ears with an operculum; above, brown spotted with white; below, white barred with reddish brown. Length 9 to 10 inches.
White's Thrush	Fourteen tail feathers. Above, buffish brown; below, white, suffused on breast with buff; marked above and below with dark crescentic bands; axillaries black with white bases. Length 12½ inches.
Black-throated Ousel	Above, olive brown; throat and breast black with pale margins to feathers; flanks grayish brown, shading into white on belly; axillaries chestnut. Female without black throat and breast, the feathers having dark centres only. Length 9½ inches.
Black-bellied Dipper	Similar to Common Dipper, but breast as well as belly black. Length 7 inches.
Bluethroat	Above, brown, except pale buff eye-

Species.	Points of Distinction.
Bluethroat (*continued*)	stripe, and basal half of tail (except two centre feathers) which is chestnut; throat and upper breast blue with chestnut spot in centre; below blue is a narrow black band, then broad band of chestnut shading into pale buff. In female the showy throat is absent, replaced by dark band across the chest. Birds of year resemble female in colour. Length 6 inches.
Desert Wheatear	General colour above buff, shading into white on rump; chin, throat, wings, axillaries, and terminal half of tail black; under parts buffish white. Female duller in colour. Length 5½ inches.
Black-throated Chat	General colour buffish white, purest on head and rump; wings, chin, upper throat, two centre tail feathers except the base, and the tips of the remaining ones, black. Female nearly uniform brown above, buff below, feathers on throat having concealed black bases. Length 5½ inches.
Isabelline Chat	Similar to female Wheatear in colour; axillaries white; terminal half of tail black, rest white. Female very similar in appearance to that of Desert Chat, but feet larger. Length 5½ inches.
Red-breasted Flycatcher	General colour olive brown; crown, nape, and sides of neck gray; tail white at base of all except two centre feathers; throat and

Species.	Points of Distinction.
Red-breasted Flycatcher (*continued*)	breast chestnut; remainder of under surface white. Female and bird of year has throat buff. Length 5 inches.
Rufous Warbler	Upper parts rufous brown; eye-stripe buffish white; tail rich chestnut, all but two centre feathers tipped with black, then white; under parts buffish white. Length 7 inches.
Yellow-browed Willow Wren	Olive green above, nearly white below; two yellow wing bars; pale stripe on crown and over each eye. Length 4 inches.
Firecrest	Similar to Goldcrest in colour, but black bands on side of crest meet on forehead; white eye-stripe; black streak through eye; crest brilliant orange in centre. Length 4 inches.
Continental Long-tailed Tit	Similar to British form, but head entirely white. Female has crown white streaked with dusky. Length 5½ inches.
Alpine Accentor	Chin and throat white spotted with brown; wing coverts tipped with white; flanks chestnut. Female similar in colour. Length 7 inches.
Nutcracker	General colour above and below chocolate brown, spotted with white except on crown and nape; under tail coverts white; wings black shot with green; tail black tipped with white. Female similar in colour. Length 14 inches.

Species.	Points of Distinction.
Waxwing	General colour vinaceous brown; long handsome crest; line from nape to bill, and entire throat black; under tail coverts chestnut; tail tipped with yellow; secondaries and sometimes a few of the tail feathers tipped with scarlet wax-like appendages. Length 8 inches.
Pallas's Gray Shrike	Above bluish gray, except rump, which is white; below white; tail graduated, four centre feathers black, rest tipped with white, increasing in extent to outermost; only one white bar on wing. Length 9 inches.
Great Gray Shrike	Similar in colour to above, but rump suffused with gray; two white bars on wing. Length 9½ inches.
Rose-coloured Pastor	Back, scapulars, rump, breast, and belly rose pink; head (crested), wings, tail, and thighs black. Length 8½ inches.
Pine Grosbeak	General colour gray; the feathers of the head, neck, back, throat, and breast margined with rose crimson; wings and tail dark brown; wing coverts, innermost secondaries, and upper tail coverts margined with white. Females have rose colour replaced by yellow; in immature males reddish orange. Length 9 inches.

Species.	Points of Distinction.
White-winged Crossbill	Above scarlet, paler below; two white bars on wings, formed by white tips to coverts. Female has scarlet replaced by yellow. Length 6 inches.
American Crossbill	Similar to preceding, but scapulars and feathers on centre of back are dark brown instead of red with dark centres.
Scarlet Rose Finch	Bill cone-shaped, with convex sides. General colour rose crimson, brightest on head, rump, throat, and breast, and palest on belly and under tail coverts; wings and tail brown, with dull crimson margins. Female dull brown, suffused with olive on back and rump; two white bars across wings. Length 5½ inches.
Eastern Bullfinch	Larger than common species; colours purer and more vivid; no red on two innermost secondaries; white streak on inner web near shaft of outside tail feather on each side. Length 6 to 7 inches.
Canary	Head yellow streaked with dark brown; back dark brown marked with lighter brown and olive; rump yellowish green; under parts yellow; flanks streaked with dark brown. Female duller and paler. Length 5 inches.
Serin	Similar to preceding; more streaked on flanks; length of tail 2 inches. Length 4½ inches.

Species.	Points of Distinction.
Mealy Redpole	Intermediate in size and colouration between Lesser Redpole and Greenland Redpole; rump grayish white streaked with brown; wing bar white. Length 5 inches.
Greenland Redpole	Breast pink; rump, under tail coverts, and flanks nearly uniform white; much more white in plumage than preceding. Length 5¼ inches.
Lapland Bunting	Hind claw straight, and longer than hind toe. Head, throat, chest, and flanks black; white streak from eye to breast; chestnut collar; rest of upper parts black with pale buff margins to feathers; wings and tail brown; two outer feathers of latter smoke-white at tip; remaining under parts white. In autumn black and rufous portions of plumage nearly concealed by pale margins. Female has the throat dull white. Length 6¼ inches.
Little Bunting.	Head and throat chestnut, with black bands on each side of crown, round ear coverts, and on side of throat; above brown streaked with black; mantle and wing coverts margined with chestnut; under parts white streaked with black on breast and flanks. Female similar in colour, but throat paler. In autumn black stripes more indistinct. Length 4¾ inches.

Species.	Points of Distinction.
Rustic Bunting	Head black, except stripe over eyes and patch on nape, which are white; rest of upper parts chestnut streaked with black, except on rump and upper tail coverts; two white wing bars; under parts white, with row of moustachial spots, and band of chestnut across breast; flanks streaked with chestnut. Female duller in colour. In autumn black feathers nearly concealed by pale margins. Length 5½ inches.
Ortolan Bunting	Bill dark red. Head, neck, and breast greenish olive, shading into yellow on cheeks and throat; remainder of under parts dull chestnut. Female duller in colour; throat streaked with brown. Length 6 inches.
Black-headed Bunting	Head and ear coverts black; tail uniform brown. Female and young, head brown streaked with darker brown; remaining colours much duller than in adult male. Length 7 inches.
Purple Martin	General colour steel blue shot with purple; concealed patch of white on sides. Length 7½ inches.
Red-throated Pipit	Similar to Meadow Pipit in colour, but more buff in tone; upper parts with dark centres; throat and breast chestnut in spring. Female less red on throat. In winter distinguished from

Species.	Points of Distinction.
Red-throated Pipit (*continued*)	Meadow Pipit by white margins to feathers of mantle; dark centres to under tail coverts. Length 5½ inches.
Richard's Pipit	Large size; long tarsus and hind claw; pure white on outer tail feathers; breast streaked. Length 7¾ inches.
Tawny Pipit	General colour above, buffish, or neutral brown, according to locality; pale buff eye-stripe; under parts uniform buff, palest on throat; white pattern on outer tail feathers. Birds of year streaked on throat and breast. Length 6½ to 7 inches.
Alpine Pipit	Hind claw straight; head and nape slate gray; under parts buff, shading into pale chestnut on breast; light pattern on outer tail feathers white; breast unstreaked. In autumn plumage closely resembles Rock Pipit, but easily distinguished by white instead of smoke-brown pattern on outer tail feathers. Length 6½ inches.
Short-toed Lark	Front toes short and curved; hind toe short and straight; no spots below. Young with faint spots on breast. Length 5½ inches.
Crested Lark	Conspicuous crest; first primary nearly as long as primary coverts; no white on outer tail feathers. Length 6¾ inches.
White-winged Lark	Bill short, arched above; secondaries white with brown bases;

Species.	Points of Distinction.
White-winged Lark (*continued*)	outer tail feather white. Length 7 inches.
Calandra Lark . .	Similar to Short-toed Lark in colour, but large black patch on each side of neck. Length 7¾ inches.
Shore Lark . . .	Tuft of black feathers on each side of crown above the eye; forehead, eye-stripe, chin, and throat yellow; lower throat and upper breast black; tail black, except two centre feathers, which are brown, and outer feathers margined on outer web with white. Female less black on crown. After autumn moult, black parts concealed by pale margins. Length 7 inches.
Needle-tailed Swift . .	Three toes in front, one behind; shafts of tail feathers prolonged into sharp spines. Length 8½ inches.
Belted Kingfisher . .	General colour above, and band across breast, bluish slate; rest of under parts white; head with bushy crest; tail barred with white. Female has chest band mottled with brown; band across belly, and the flanks chestnut. Length 12 inches.
Yellow-billed Cuckoo .	Ten tail feathers; above, including two centre tail feathers, brown shot with green; remaining tail feathers black, tipped with white; primaries suffused with chestnut; under parts white. Bill yellow, broadly tipped on upper man-

Species.	Points of Distinction.
Yellow-billed Cuckoo (*continued*)	dible, and narrowly tipped on lower one with black. Young resemble adults, but have no chestnut on primaries. Length 12 inches.
Purple Heron.	Smaller than Common Heron, but toes longer. Crown and crest black; dorsal plumes, under wing coverts, thighs, and breast chestnut. Length 31 inches.
American Bittern	Resembles Common Bittern, but smaller; vermiculations finer; primaries uniform in colour, not barred. Length 27 inches.
Glossy Ibis	Bill curved as in Curlew; face bare of feathers; general colour chestnut, shot with green on back. Length 22 inches.
Macqueen's Bustard	Similar in colouration to Little Bustard, but vermiculations finer, and extending over wing coverts, head, and throat; crown and nape plumes slate gray, tipped with black; upper neck plumes black, lower ditto white. Female similar to male in colour, but smaller. Length 28 to 30 inches.
Cream-coloured Courser.	General colour buff; under tail coverts white; head gray; primaries, under wing coverts, and axillaries black. Length 10 inches.
Gray Phalarope	Summer plumage: under parts chestnut; axillaries and under wing coverts white. Female more brilliant in colour than

Species.	Points of Distinction.
Gray Phalarope (*continued*)	male. Winter plumage: above slate gray; under parts white. Feet lobed. Length 8 inches.
Asiatic Golden Plover	Similar to Golden Plover, but axillaries smoke-brown instead of white. Length 9 inches.
American Golden Plover.	Larger than preceding (wing 6·8 to 7·5 inches against 6·0 to 6·7 in.)
Killdeer Plover. . . .	Similar to Ringed Plover; but two black bands across chest; rump and upper tail coverts chestnut; tail long and wedge-shaped. Length 10 inches.
Esquimaux Curlew . .	Similar in general colouration to Whimbrel, but much smaller; no white on rump; axillaries chestnut barred with brown. Length 14 inches.
Bartram's Sandpiper .	Similar in general colouration to female Ruff, but quills and tail feathers barred with black. Length 12 inches.
Solitary Sandpiper . .	Similar in general colouration to Green Sandpiper, but no white on rump, which is same colour as back. Length 9½ inches.
Spotted Sandpiper . .	Similar in general colouration to Common Sandpiper, but under parts in summer spotted with black. Distinguished from Common Sandpiper at all ages and seasons by having more brown than white on secondaries; white on outer tail feathers is less than in its Old World ally. Length 7½ inches.

Species.	Points of Distinction.
Yellow-legged Sandpiper.	Similar in general colouration to Wood Sandpiper, but larger, and legs proportionately longer (2 inches, as against 1½ in Wood Sandpiper), and bright yellow instead of pale olive; less white on rump, and obscurely barred at all ages. Length 10¾ inches.
Red-breasted Snipe.	No web between middle and inner toe; bill Snipe-like. Under parts in summer buffish chestnut; white in winter; upper parts gray in winter; rump white barred with dark brown. Length 8½ inches.
Bonaparte's Sandpiper.	Similar in general colouration to Little Stint, but always distinguished by having upper tail coverts pure white, streaked sparingly with dusky brown. Length 7½ inches.
Pectoral Sandpiper.	Similar in general colouration to Little Stint, but larger; breast gray, thickly streaked with dark brown in summer, and in birds in first plumage; in winter only sparingly streaked; legs and feet pale buff. Length 8½ inches.
American Stint.	Generally smaller than Little Stint (length of wing 3·3 to 3·6 inches; Little Stint 3·6 to 4·0 inches); other characters very unstable; of doubtful specific rank. Length 6 inches.

Species.	Points of Distinction.
Buff-breasted Sandpiper	Similar in general colouration to Bartram's Sandpiper. Tail graduated; all (except two centre feathers) tail feathers, under surface of wings, greater and primary wing coverts, buff, marbled with black and white. Length 7 inches.
Sabine's Gull	Tail forked; head dark slate gray; black collar round neck. Hood absent in winter. Length 13 inches.
Ross's Gull	Tail long and graduated; head rosy; collar black. Length 14 inches.
Glaucus Gull	General colour white; mantle, scapulars, and wing coverts gray. Head and neck streaked with gray in winter; orbits vermilion. Young have primaries grayish white. Length 28 to 33 inches.
Iceland Gull	Similar to preceding in colour, but smaller; adult's orbits flesh colour. Length 22 inches.
Ivory Gull	Entirely white; orbits vermilion. Immature birds have black spots on wing coverts, and black tips to primaries and tail feathers; still younger birds have black spots also on mantle and scapulars. Length 16 to 18 inches.
Little Gull	Smallest Gull (wing 9½ inches). Head black in summer, white in winter streaked with gray. Length 10 to 11 inches.

Species.	Points of Distinction.
Pomarine Skua	Tail broad and rounded in adults; centre tail feathers long and twisted (4 inches longer than rest); tarsus longer than middle toe and claw. Length 21 inches.
Buffon's Skua	Central tail feathers long and pointed (9 inches longer than rest); first and second primaries only with white shafts; nostrils nearer frontal feathers than tip of bill. Length 22½ inches.
Little Auk	Resembles Razorbill in general appearance, but very small. White spot over eye; chin, throat, and fore neck white in winter. Length 8½ inches.
White-billed Diver	Similar to Great Northern Diver in colour, but bill stouter, curved upwards, and pale yellow; six white streaks on front of throat; ten on sides of neck. Length 34 inches.
Sooty Shearwater	Underparts uniform brown. Length 18 inches.
Bewick's Swan	Smaller than Hooper; yellow on base of bill not extending below nostrils. Length 50 inches.
Snow Geese	Entire plumage in adults white, except primaries, which are black. Length of wing 15 to 18½ inches. Bill, legs, and feet red. Young, pale slate gray, darkest on scapulars and wing coverts, which have pale margins.

Species.	Points of Distinction.
Red-breasted Goose	Forehead black; throat and breast chestnut; white spot between eye and bill. Young have chestnut and black parts of plumage paler and browner; pale margins to wing coverts. Length 22 inches.
Ruddy Sheldrake	General colour buff, palest on crown; wing coverts white; speculum green; black ring round neck. In female this ring is absent. Length 25 inches.
Red-crested Pochard	Crested; head and neck bright chestnut; speculum and axillaries white; bill and legs vermilion. Female uncrested, crown dark brown; cheeks, neck, and sides of throat gray; bill and legs reddish brown. Length 21 inches.
American Teal	Broad white crescent on each side of breast in front of shoulder. Females indistinguishable from female Common Teal. Length 13 to 15 inches.
American Wigeon	Broad green stripe on side of head, from eye to neck; axillaries white; ground colour of back and flanks vinaceous. Female, black speculum, shot with green; axillaries white. Length 20 inches.
Buffel-headed Duck	Similar to Golden Eye, but smaller, and the white patch on side of head behind the eye instead of before it. Female with similar white patch. Length 14 to 15 inches.

Species.	Points of Distinction.
Surf Scoter	Wing and scapulars black; axillaries brown; white patch on crown, and another on hind neck. Female duller in colour, and with only white patch on nape; axillaries brown. Length 21 inches.
Steller's Eider	Black ring round neck, shot with purple and green; black ring round eye; two black spots on sides of breast; lores and nape emerald green; speculum blue, shot with purple. Female distinguished by metallic speculum between two alar bars. Length 18 to 20 inches.
Hooded Merganser	Crest well developed, white bordered with black; head and neck black. Female has head and neck dark brown. Bill narrow and furnished with saw-like lamellæ. Length 17½ inches.

CHAPTER IV.

MIXED CONGREGATIONS.

HE who pays attention to the habits and movements of birds during the months of autumn, will not fail to remark the great increase of their gregarious and social instincts. Birds seem compelled to fraternise, and even the most solitary ones show an inclination to join with their own kind into parties, or to attach themselves to flocks of very different species. It may be that many of these gatherings are for the purpose of increased safety, a flock being always more difficult to approach than single birds. During the period of their migrations, birds seem to love companionship, and not only travel with their own kind, but with birds of many other species. Many young birds are even more gregariously inclined than their parents, and as soon as ever they are able to take care of themselves, they form into parties and flocks for the remainder of the year. Old birds do not display much social tendency until after they have completed their autumn moult, and some species remain solitary after this event is over much longer than others. It is also worthy of remark how many of the accidental visitors to this country have been shot out of flocks of very different species : the Red-breasted Goose among the Brent and Bernacle Geese, and the White-winged Lark among the Snow Buntings may be cited as cases in point.

I know of few things more pleasing at this time of year than to stand and watch the movements of these mixed congregations of bird life, say, as the sun goes down behind the distant hills, setting the painted woods aflame, and glinting on the branches of the forest monarchs, causing the bark of the fir trees to glow like purple fire; or on the moors or mud-flats at break of day. Ah, those autumn mornings and evenings! How I love them! What stirring scenes among the birds I can recall; what stores of notes I made! How vividly some of them return; incidents of twenty years ago, among what were peaceful fields and wooded valleys then, but now, alas, the busy centre of a score industries. All those mighty trees, where the Rooks and Ring Doves bred, have bowed their noble heads before the axe; all the brushwood and the thickets and close-set hedgerows, where the Warblers and Finches nested, are gone; all the tangled ditches and hollows, sacred to the Grasshopper Lark, and the Jack Snipe, levelled away! Reader, pardon this digression; but to write on bird congregations is to bring back from the past a whole train of memories concerning one of the most highly-favoured spots for birds it has ever been my lot to know. Sad to relate, this bird paradise has all been destroyed; misfortunes and death overtook the ancient family who owned this fair domain, and fields and park, woods and fish-ponds, all fell a prey to that modern curse, the speculative builder! No longer does the air resound with song at morn and even; all the feathered hosts are gone; the trout-stream is little more than an open drain; and one of the fairest sylvan scenes that ever eye of man gazed upon is now a desolation of bricks and mortar, and a wilderness of tall chimney-shafts, factories, and workshops! My ruined aviary! No other rural spot has ever yet been able to console me for its loss. I knew every tree and bush, and bird and beast within it, and loved them all!

But to return to the birds. It was here that every year I used to watch the autumn flights of Finches—a mixed and merry congregation in the beech woods and on the open fields. No two species are more regularly found in company than the Brambling and the Chaffinch; and very often a fair sprinkling of Greenfinches and Yellow Buntings will join the merry, restless throng. It is a pretty, animated sight—four of the gayest of our Finches all in company, flitting about among the undergrowth, or settling in a noisy throng upon the tree-tops, whence they descend, one after the other, to the fields below, to pick up the scattered seeds. At eventide the three species of true Finch will most probably roost in company; but the Buntings will just as likely retire to a separate locality for that purpose. Other mixed congregations of bird life may often be met with in the autumn woods. It will almost invariably be remarked that the birds which fraternise together are very similar in their habits and movements, and subsist on much the same kind of food. When the autumn is waning, and the birch and alder trees have lost much of their painted foliage, we may be sure to meet with parties of Goldcrests among them. These little birds are almost invariably accompanied by a few Coal Tits; and sometimes a brood of Blue Tits will join the picnic. Now and then the Tits are the most numerous, and only a pair or so of Goldcrests accompany them; whilst, on rarer occasions still, a flock of Siskins, and more frequently a party of Lesser Redpoles, will join the general congregation. The seeds on the alder and birch trees are a great attraction; and we may find amusement by the hour together in watching the various attitudes assumed by these active little birds in searching for their food. From tree to tree fly the eager birds, from one end of the grove to the other; then back again, one after the other — Tits, Goldcrests, and Siskins, all mixed up together in the greatest confusion, yet readily distinguished by their various notes. Another mixed party may often be seen in the woods in late autumn, composed of a Nuthatch, a pair of Creepers, and occasionally a Lesser Spotted Woodpecker. All are bent on the same purpose—to search the nooks and crannies of the bark, and the branches and twigs for insects. Up and down the trees the merry little party go, twittering to each other, apparently to disclose their whereabouts, and thus keep together. Rooks, Starlings, Redwings, and Thrushes also often form a dinner-party on the fields; but the Thrushes are only attracted by the food, the Rooks and Starlings alone showing any desire to be sociable.

Many mixed congregations gather together on the purple expanses of moorland in autumn. Birds form here into flocks and parties, to feed upon the sumptuous banquet of wild fruits, such as bilberries, cranberries, and blackberries. All the birds of the moor are fond of these berries, and many different species may be seen devouring them in company. I have a note to the effect that on the south Yorkshire moors, in the neighbourhood of the Peak, a flock of Ring Ousels, numbers of Red Grouse, several Missel-thrushes and Stock Doves, besides numerous Meadow Pipits, and a pair of Lapwings, were all flushed from one small patch of bilberry and heather not more than a couple of acres in extent. Again, it is worthy of note how this propensity for becoming gregarious and sociable is entirely absent from many species. None of the Warblers, for instance, can be called gregarious or very sociable. Rooks, Jackdaws, and Choughs are fond of society; Magpies and Jays like to live solitary or in pairs. Nearly all the Finches, the Tits, Buntings and Larks, Wagtails and Pipits become gregarious in autumn, and more or less sociable with other and often distantly related species.

Mixed congregations of quite a different character may frequently be observed on the sea-coast at this season; some of them on the mud-flats, for instance, are intensely interesting. From far and wide these little strangers come—Knots and Dunlins from the Arctic regions; Golden Plovers from the moors; Black-headed Gulls from distant inland breeding-stations; wary Herons from the adjoining woods; Oystercatchers from the rock-bound coasts further north, all assembled here to search the gleaming sands and muds for food, as soon as ever the tide has ebbed. Far out on the open banks the first small flocks of Geese have assembled — perhaps three or four species in every company—a noisy, restless host, some busy preening their feathers, others fast asleep. Flocks of young Knots mingle indiscriminately with Dunlins, and stray examples of Purple Sandpipers, Sanderlings, Curlew Sandpipers, and Stints, lost and separated from their companions, join the general throng, apparently for company's sake. Few birds are more sociably or gregariously inclined than these little sand birds. Sometimes we may meet with a solitary bird upon the muds, but it is almost invariably a sickly or a wounded one, too weak to keep up with the rest. Very often a mixed gathering of Ducks and other water-fowl may be met with in the late autumn days on some secluded pool where the shy birds know they are safe from danger. By parting the tall, breast-high rushes aside, and peeping cautiously through them, you may often watch the movements of these swimmers unawares. There are the shy Wigeon, easily identified hundreds of yards across the water by their pale yellow foreheads; gaudy Mallards, in all the splendour of their wedding-dress, float and paddle side by side with Pochards and Tufted Ducks, the former easily recognised by its dark chestnut head and neck, and the latter by its handsome crest. Charming little Teals swim round and round, and in and out of the tufts of dead reeds; and now and then a Coot or a Moorhen joins the party, and paddles away again as if unsatisfied with its welcome. Every few minutes the Pochards and Tufted Ducks dive with marvellous speed, disappearing under the water as if by magic, then coming up again with pieces of waterweed which they have pulled up from the bottom of the lake. The Mallards and other non-diving Ducks seek their food in the shallows, sifting the fine mud, and poking and dabbling everywhere round the shore. Ducks are sociable birds, and never seem happy alone; right through the autumn and winter they live more or less in flocks, feeding, flying, and sleeping in company.

Coots and Moorhens gather in autumn, especially on the broads; and sometimes the various species of Grebes will be found in their company. The Gulls and Terns are also sociably inclined, and their congregations are often very mixed in character, though in this group of birds the young ones keep much aloof from the adults, and the smaller species do not flock much with their larger relatives. These birds, however, are perhaps most gregarious and sociable during the breeding season, when Gulls of many species rear their young side by side, as we have already seen.

The vast flights of Ring Doves and Stock Doves, which congregate with the farmers' Pigeons on the stubbles in autumn, form a mixed congregation that always affords me pleasure. Pigeons are remarkably sociable birds, and when once the brown tints of autumn tell us that the year has turned, these birds unite into companies for the winter. Flock after flock come from the surrounding woods, especially from those where fir trees are numerous, and meet at the common rendezvous. Sometimes these birds gather in the woods among the oaks and beeches, where the acorns and the mast are the attraction, the branches resounding with the rattle of their wings as the birds fly from tree to tree. I often see the Ring Doves and the Stock Doves congregated in the woods at places where the Pheasants are fed, all three species fraternising most amicably. These two species of Pigeon are more or less sociable with the Rock Dove in all districts near the sea.

Such are a few of the mixed congregations to be seen in various districts during the autumn. Many others may be met with throughout the haunts of birds, which will not fail to interest the observer. Birds are even more attractive and engaging when massed in flocks, than when living alone, and, what is very important to the naturalist, the habits of most species are considerably changed during these gregarious and social periods.

CHAPTER V.

WHERE THE MIGRANTS GO.

ALL through the autumn we have been watching the gradual departure of birds from their summer haunts. By the time November's cold and cheerless month arrives, but very few indeed of these summer birds are left; they have all sped southwards with the sun. Let us devote a few pages to these little migrants after they have left their northern home; let us follow them south and learn the secret of their destination. Many there are that go to the remotest regions of Africa; many more that go only half that distance. We may conveniently divide the winter quarters of our migratory British birds into three fairly well-defined districts or zones. The first of these embraces the south of Europe, and the narrow border of fertile country along the African coast-line, which may be expressed as the Basin of the Mediterranean; the second zone consists of the Great Desert, and the entire Soudan, probably as far south as the equator; whilst the third zone extends from the equator to Cape Colony. It must, however, be borne in mind that many of the birds from North-East Europe winter in Asia Minor, Arabia, Persia, and Turkestan, whilst a considerable contingent of birds from West Siberia find their way south-west to Africa. It is highly probable, however, that every one of our British summer migrants either winters in the basin of the Mediterranean or in the two lower zones of Africa. Some of them inhabit South Europe as well as the middle zone of Africa, and others do not cross the Mediterranean at all.

It is rather noteworthy how many of the British Birds of Prey are migratory—" British" implying those that undoubtedly breed in our islands. The two greatest travellers are the Hobby and Montagu's Harrier. Both these birds are excessively rare with us, yet their migrations are as regular and as interesting as those of much commoner species, whose annual movements are better understood. The winter home of these two birds is in the lower zone of Africa, extending to the Cape Colony. The species that winter not quite so far away from us—that is to say, in the country between the equator and the Atlas Mountains—are the Kestrel, the Honey Buzzard, and the Marsh Harrier; whilst those that travel least of all are the Osprey and the Hen Harrier, which find a suitable winter haunt in the basin of the Mediterranean, many individuals remaining on the northern shores of that sea. The Ring Ousel is the only other migrant the British examples of

R

which winter exclusively in the basin of the Mediterranean, both in South Europe and North Africa, but do not descend beyond the Atlas into the oases of the Great Desert.

By far the greater number of our smaller migrants find their winter home in the country lying between the southern slopes of the Atlas and the equator. Some of the birds that retire to this central zone winter in localities as yet unknown to naturalists. Where the Nightingale, the House Martin, the British examples of the Sand Martin, and the Turtle Dove spend the time that they are away from us remains to be discovered; it is some part of the Central Soudan, and in the very remotest oases of the Great Desert, where no white man has ever yet penetrated. There are several birds that not only winter in the Mediterranean basin, but extend their wanderings into the middle zone, which reaches to the equator. These are the Whinchat, among the Turdinæ; the Grasshopper Warbler, the Blackcap, the Willow Wren, and the Chiffchaff among the Sylviinæ; and the Tree Pipit, among the Motacillinæ. All these species winter in more or less abundance in the extreme south of Europe, as well as in the African Continent north of the equator. In all these birds, with the exception of the Whinchat and the Tree Pipit,* the

---

* A few Whinchats and Tree Pipits breed on the mountains of South Europe, in a similar climate to that of their usual habitat further north.

WHERE THE MIGRANTS GO.

winter range overlaps that of summer; although it is most probable that the individuals breeding in this winter area move south in autumn, and are replaced by others breeding farther north. Thus, the Chiffchaffs and the Willow Wrens, that breed in South Europe and North Africa, winter in the Soudan, and those that breed in North and Central Europe take their places for the winter. There is as yet no evidence to show that any individuals of these species are resident, or remain stationary in any part of their range. One or two species winter in all three zones, but a few individuals remain behind to breed in the extreme north of Africa, as well as in the south of Europe. These are the Wryneck, the Quail, the Corn-Crake, the Ringed and Kentish Plovers, and the Common Sandpiper.

We now come to those birds that winter exclusively in the middle zone, none of them remaining in Europe at that season, or penetrating, as a rule, below the equator. Some of these winter homes of our feathered friends are indescribably beautiful—the fair oases in the Great Desert, where food is plentiful, and the weather genial. Here, among the waving date palms and pomegranates, the tamarinds, oranges, and oleanders, the Redstart, the Pied Flycatcher, and the Wood Wren chase the flies among the foliage just as happily as in the English coppices; the Wheatear, all northern lands forgot, fraternises with the Chats and the Larks of the desert; the Reed

Warbler and the Marsh Warbler love to skulk among the dense vegetation by the pools, and catch the insects that dance in myriads in the brilliant sunshine; and in the half-dry beds of the desert streams, the pretty Dotterel, tame and confiding as ever, runs up and down the sand. With the dawn of spring, most of these birds hurry away from this land of promise; when the sun reaches the northern tropic, everything down here will be scorched and parched, the waters will dry up, and all will become deserted and desolate. In the north, however, tempted by the beautiful climate of the Tell, a few birds will remain behind to rear their broods after their companions are gone. These are the Pied Flycatcher and the Wood Wren.

Although this district is the grand winter home, not only of our own summer birds, but of the countless millions that inhabit the whole of the northern parts of the west Palæartic region, the remaining half of the hot and sultry African Continent is a favourite wintering place of a great many species. Some few there are that may be found in winter south of the Atlas, from one end of Africa to the other. These are the Spotted Flycatcher, the Sedge Warbler, the Garden Warbler, and the Common Whitethroat, the Yellow Wagtail, the Swift, and the Goatsucker. Some few of these remain behind to breed in the extreme north of Africa, as, for instance, the Spotted Flycatcher, the Swift, and the Goatsucker. I have seen the Spotted Flycatcher literally in swarms in the more northern oases during May. It seemed as though all the Spotted Flycatchers of Europe were congregated here, loitering away the last few days of their stay in Africa previous to migrating north. The birds, however, that make the southern half of Africa their home in winter are comparatively few. These are the Swallow, the Red-backed Shrike, and the Cuckoo. A few Swallows are said to winter in some of the oases in the Sahara, but such instances are exceptions to the rule. The migrations of the Red-backed Shrike are very interesting, inasmuch that this bird sets off from England in a south-easterly direction, instead of due south, crossing Europe and Asia Minor, and passing down the valley of the Nile, across the Lake Region and the Zambesi to South Africa. Few, if any, individuals of this species travel by the much shorter western route, as the Red-backed Shrike is rarely, if ever, seen in the Spanish peninsula, or along the West African coast. A few Cuckoos remain behind to breed in the rich and fertile districts of Algeria, but it is not known that any of these individuals remain north of the equator during the winter. It remains now but to notice the Terns. All the British species are migratory, and have their winter home in the warm seas round the African coast, and the Lesser Tern spends that season in some numbers on the Nile. Doubtless these individuals are the birds that breed in Persia and Turkestan, as a very important stream of migration drains much of Western Asia through these countries down the Nile valley to Central and South Africa.

Many birds pass along our coasts in autumn, on their way south, in sufficient abundance to render them familiar to the British naturalist. One of the most noteworthy of these is the Rough-legged Buzzard, on its way to the south of Europe, but it is not known to cross the Mediterranean into Africa. Another of these birds is the Knot, numbers of which remain on our coasts to winter, but the great majority pass on to the western shores of Africa. In the same way, numerous Geese and Ducks give our coasts a call (many remaining for the winter) as they pass on to more southern latitudes.

The routes taken by our summer migrants are followed with great regularity and persistence. Those birds wintering in Morocco, West Africa, the Congo and Damara Land appear to enter Africa by way of Gibraltar, reaching that locality from the southern shores of England, by way of the Channel Islands to Finisterre, or along the Seine and on through Bordeaux, across the Pyrenees, and down the Portuguese coast. Those wintering further east in South and Central Africa, in Algeria, Tripoli, and Egypt, cross the English Channel and fly over France, down the great river valleys of Central Europe, through the passes of the Alps, and along the coasts of Italy and Turkey, crossing the Mediterranean at certain well-adapted points, as by way of the Balearic Islands, Corsica, Sardinia, Sicily, Candia, the Greek Archipelago, and Cyprus. These routes make the actual distance traversed across the sea comparatively small; and probably, at the vast height these little travellers fly, land is rarely if ever lost sight of.

The habits of our summer migrants have been but little observed in the southern lands to which these birds retire in autumn. So far as we can ascertain, some of them regain their lost songs and warble through the winter. Such species as the Willow Wren, the Chiffchaff, and the Blackcap sometimes sing before they leave this country in autumn, and the latter bird is known to warble all the winter through, if we can apply the term "winter" to a land of almost perpetual summer. But though these birds may engage in song, just like the Robin and the Wren in our northern woods, there is not a scrap of evidence to show that any of them breed again in their winter quarters. None of our summer migrants ever appear to breed south of the equator, and of those that do breed in Africa, they only nest during the period of spring and summer in the northern hemisphere. With the exception of the Swallows and the Red-backed Shrike, all our summer migrants moult before they leave us in autumn; and all the species which have a spring moult (including the Swallows and the Shrikes) complete their change of dress previous to setting out on their northern journey. The sexes of many species appear to separate for the purpose of migration, though whether there is any difference in the locality chosen for wintering, or in the routes followed, yet remains to be determined.

From these particulars of the various destinations of our summer birds, it will be seen that the vague expression "winters in Africa," so often and generally applied to these little creatures, by no means represents the philosophy of their annual movements, and that a study of the winter distribution of our feathered favourites is replete with unusual interest, and forms by no means the least charming portion of their economy.

CHAPTER VI.

PARAGRAPHS ON PLUMAGE.

THERE is no more appropriate season than the present for a few remarks on the plumage of birds. Questions relating to the various changes and characteristics of plumage are constantly coming before the notice of the student in autumn, for this is the season when every bird of the northern hemisphere renews its worn and faded dress. The dull-plumaged young birds acquire the brighter and more varied tints peculiar to their parents, or assume a winter dress like the adults. To enter into the colour details of birds' plumage may seem a dry and uninviting task, yet the subject is of great importance and is surrounded with no ordinary degree of interest.

The student of birds will soon become cognisant of the fact that various peculiarities of plumage are either common to certain well-defined groups of species, or are present in many birds in which relationship is of the remotest kind. These peculiarities are known to naturalists as "patterns of colour," and are often more reliable and constant characters for separating species into groups or genera than many structural ones. Pattern of colour is infinitely less liable to variation than any structural character. Among the Passeres especially, the systematic naturalist is often at his wits' end to give a point of distinction which will separate a number of obviously allied birds into a group or genus. But when structural characters fail, he is often helped out of his difficulties in a very remarkable manner by the pattern of colour on the plumage which extends through an entire group of species.

Some of the most curious instances of this generic pattern of colour exist in the TURDINÆ, which includes the Thrushes and their allied forms. The Ground Thrushes, of which the well-known White's Thrush is a typical species, are separated from all other allied groups by the pattern of colour on the under surface of the wing, formed by the basal half of the outer web of the secondaries and some of the primaries, which is either pure white or light buff. This singular and characteristic pattern extends through all the forty species and races which comprise the genus *Geocichla*. The typical Thrushes (numbering some fifty species), of which the Missel-thrush is a good example, are readily distinguished from the Ousels or Blackbirds and the Ground Thrushes by the streaked throat and the absence of the wing pattern. In this group the sexes are alike in colour. The Blackbirds are separated either by their unstreaked throats, or the important and marked difference of colour between the sexes—characters which embrace the fifty species of which this group is composed. The dozen species of Redstarts are characterised by the colour of their throat (black or dark blue), which is in violent contrast with the colour of the breast, and by the rich chestnut of all but the two central tail feathers of the majority of species. Most of the Chats (of which our Common Wheatear is a typical species) are distinguished from their relations by their white rump, upper tail coverts, and basal portion of the tail feathers. Another very pretty generic pattern of colour is presented in the Jays, numbering a dozen species, in all of which the wing coverts are barred with black, white, and blue. These patterns of colour are most probably of very remote origin, have been transmitted in a scarcely varying way from the earliest history of the species, and have been preserved while many structural characters of those species have undergone important changes.

The manner in which certain patterns of colour are common to distantly related species is also very wonderful and interesting. It is also worthy of remark that these colours are confined to certain parts of the plumage only. What a great variety of birds, for instance, exhibit the character of a black cap or hood! Many of the typical Warblers are so marked, many of the Titmice, many of the smaller Gulls, the

Terns, and many of the Bullfinches; whilst in numbers of species the cap is brown or some other shade, in violent contrast to the colour of the rest of the plumage. Then there are others in which the peculiar marking is confined to the chin, as is displayed in such a marked degree by the Hawfinch, the Waxwing, and the Redpole; whilst in scores of others this dark colour extends over the entire throat, among which may be mentioned the Stonechat, the Black-throated and Desert Wheatears, and the Reed Bunting. The plumage on the throat and breasts of birds appears to have been specially selected for the stamping of various patterns in exceptionally brilliant colours, or ones in strong contrast to the remainder of the plumage. How many birds there are whose most brilliant hues or most striking characteristics are carried on the throat and breast! Among them may be mentioned the speckled Thrushes and Pipits, the Bluethroats, the Robin, and the Rock Partridges.

Before leaving the heads of birds, we must say a few words on crests. These curious and beautiful appendages are common to almost every class of bird — from the Eagles to the Titmice. The charming Waxwing and the crested Titmouse, the crested Lark and the Hoopoe, possess these ornamental head-dresses in common with the Lapwing and the Grebes. Crests, too, have been selected for the display of considerable variety and beauty of colour, some of them being brilliantly painted in the most gorgeous tints. Some birds have a few hair-like filaments at the back of the head, only detected by the closest scrutiny — the Blackbird, for example.

On many species the chief pattern of colour is stamped on the belly. This is most remarkable on birds of otherwise protectively coloured plumage, which depend for their safety on the harmony of their colours with surrounding objects. A familiar instance is to be seen in the brown horse-shoe mark on the Common Partridge, and the brilliant black belly of the Golden Plover and the Dunlin in breeding plumage. Others have this colour pattern on the lower back and upper tail coverts. In how many birds, for instance, does a white rump prevail—species, too, as widely distinct as the Rock Dove and the House Martin, the Wheatear and the Bullfinch, and some Petrels. The mantle is also another portion of a bird's plumage, which is often emphasized by the same colour in many remotely allied species, or in entire groups of others. The region of the vent and under tail coverts has also been selected for the display of much brilliant pattern of colour, as in the Bulbuls, for example.

The wings of birds have also much pattern of colour in common. In a great number of species the wing coverts are selected for the display of various brilliant or striking colours; in others, bars are exhibited; in some the quills are particoloured; in many all the outer margins of the flight feathers are edged with brilliant tints, even in the young, as is the case with the Goldfinch. The tail has been equally modified. There are species among almost every great group of birds in which the tail feathers are barred or marked in certain well-defined patterns. In a great many cases the bars are replaced by tips of a different colour from the remainder of the feathers; whilst in other cases these bars are broken up into well-defined spots, which extend across the entire tail.

As yet we are utterly in the dark respecting this singular coincidence in the pattern of colour on the plumage of so many widely different species. It would seem as if the various internal organs of birds were emphasized in many cases on their dermal covering; and that for some reason, which still remains to be discovered, certain parts of the body are more likely to produce these colour patterns than others. The head, the back, the throat, the breast, the belly, the vent, the wings, and the tail have each and all been marked on the plumage in a variety of brilliant or striking colours. It seems as if these portions of the plumage were singularly sensitive to the development of such colours, perhaps aided by Sexual Selection and Isolation.

A word now as to the texture of plumage. This, like everything else in nature, is wonderfully adapted to the habits and requirements of the birds possessing it. Owls are birds of singularly soft and fluffy plumage, well adapted to their nocturnal habits, when these birds have to pounce softly and silently on their prey at a time when the least sound is heard far and wide. Yet the Goatsuckers also exhibit this peculiarity, which has probably been inherited from the common ancestor of the Owls and the Nightjars. The plumage of some birds is remarkably loose and pervious, as in many of the Passeres; hard and compact, as in the Pigeons; dense and waterproof, as in the Ducks and sea-birds. Birds which inhabit cold climates are well provided with a warm covering, two of the most remarkable instances being the Siberian Jay and the Lapp Titmouse, which are constant residents in forests where the temperature in winter is so low that few creatures can exist.

The whole subject of birds' plumage has been little studied or investigated, yet it presents an unusually rich field for careful observation and patient research. I cannot but think that the investigation, if properly made, will throw much light on the past history of birds, and clear up many points in their affinities which are still buried in clouds of obscurity. I offer these few remarks merely as hints of encouragement for the reader to take up the study of this branch of ornithology, rather than any serious discussion of the question, which is too complicated and far too important to be dealt with in a few cursory pages.

CALENDAR FOR AUTUMN.

Species.	September.	October.	November.
Hobby.	Migrating	Migrating	Absent
Merlin	Young move South	Old birds on lowlands	Occasionally seen
Osprey.	Migrating	Migrating	Absent
Honey Buzzard . .	,,	,,	,,
Hen Harrier . . .	,,	,,	Occasionally seen
Montagu's Harrier .	,,	,,	,,
Short-eared Owl . .	Moulting	,,	Migrating
Missel-thrush. . .	,,	Some migrants arrive	Large flocks disperse
Song Thrush . . .	On turnip fields	Migrants passing	Many migrating
Redwing	—	Arrives	Flocks in cultivated districts
Fieldfare	—	Begins to arrive	Still arriving
Blackbird	Common in fruit gardens	Grass lands morn and evening	Many migrating
Ring Ousel . . .	Leave moors	Migrating south	Absent
Robin	In full song again	Young wander	Leave exposed districts
Nightingale . . .	Migrating	Absent	Absent
Redstart	,,	,,	,,
Black Redstart . .	—	Arrives	Still arriving
Wheatear	Migrating	Absent	Absent
Whinchat	,,	,,	,,
Stonechat	Decreasing in numbers	Leaves higher ground	Seen on lowlands
Spotted Flycatcher .	Migrating	Absent	Absent
Pied Flycatcher . .	,,	,,	,,

CALENDAR FOR AUTUMN.

Species.	September.	October.	November.
Grasshopper Warbler	Migrating	Absent	Absent
Sedge Warbler . .	,,	Migrating	,,
Reed Warbler . .	,,	,,	,,
Blackcap	,,	,,	Occasionally seen
Garden Warbler . .	,,	,,	Absent
Whitethroats . . .	,,	,,	Occasionally
Chiffchaff	,,	,,	,, [seen
Willow Wren . . .	,,	Absent	Absent
Wood Wren . . .	,,	,,	,,
Goldcrest	In parties	Migrants arrive	Go South, or disperse
Titmice	,,	Wander far and wide	Not so gregarious
Hedge Accentor .	Regains song	In turnip fields	Near houses
Wren	In song again	Full of music	Leaves exposed dist'cts. during severe weather
Creeper	Broods disbanded	Often seen with Tits	In company with Parinæ
Nuthatch	Solitary	Feeds on beech and other nuts	Food chiefly insects
Hooded Crow . .	Completes moult	Migrants arrive	Live on coast and follow tidal rivers
Rook	,,	Visit old nests daily	Exceedingly gregarious
Jay	,,	Parties disbanded	Live in woods
Red-backed Shrike.	Migrating	—	Has been seen
Starling	In flocks	In flocks	In flocks
Bullfinch	Completes moult	Wanders in pairs	Wanders in pairs

s

Species.	September.	October.	November.
Hawfinch	Completes moult	Feeds on beech-mast	Still in parties
House Sparrow . .	,,	In flocks on stubbles	Draws nearer homesteads
Greenfinch . . .	Few pairs still nesting	In flocks	Desert shrubberies in daytime
Goldfinch	Many migrants passing	Many migrants passing	In small flocks
Siskin	,,	,,	,,
Brambling . . .	Absent	Absent	Arrives
Chaffinch	Completes moult	Migrants arrive	In flocks
Linnet	,,	,,	,,
Twite	,,	On stubbles	,,
Lesser Redpole . .	Completes moult	In flocks	Often on clover stubbles
Snow Bunting . .	Migrating in small numbers	Flocks arrive	Chiefly frequent coast
Cirl Bunting . . .	Completes moult	Flocks with Chaffinches	On stubbles in S.W.
Yellow Bunting . .	,,	Gregarious	On stubbles and newly sown fields
Pied Wagtail . . .	Many move South	Still passing South	On ploughed lands
Gray Wagtail . . .	Quits uplands	Near lowland waters	—
Yellow Wagtail . .	Migrating	Migrating	[seen Occasionally
Meadow Pipit . .	Numbers passing along our coasts	Vast numbers passing	Lives on lowlands
Tree Pipit	Migration begins	Migrating	Absent

Species.	September.	October.	November.
Rock Pipit	In flocks	Many arrive	On low-lying coasts
Skylark	Migrants arrive	Millions passing	Migration continues
Wood Lark	In parties	Roaming about	Roaming about
Swallow	Migration begins	Migrating	Absent
Martin	,,	,,	,,
Sand Martin	Migrating	,,	,,
Swift	,,	Absent	,,
Goatsucker	,,	,,	,,
Wryneck	,,	,,	,,
Ring Dove	Silent	Many migrants arrive	On stubbles and new-sown land
Stock Dove	,,	Eggs occasionally found	,,
Turtle Dove	Migrating	Occasionally seen	Occasionally seen
Ptarmigan	Moulting	Rapidly assuming white dress	—
Red Grouse	In large packs	Resort to higher ground	—
Partridge	Frequent grass fields	Flocks become larger	—
Quail	Migrating	Migrating	Occasionally seen
Heron	Frequents mud-flats	Lives on mud-flats	Roaming about
Corn Crake	Migrating	Absent	Absent
Moorhen	Young leave birthplace	Feeds on berries	Roosts occasionally in woods

Species.	September.	October.	November.
Coot	Draws near to coast	Many migrants arrive	Very gregarious
Stone Curlew	Completes moult	Leaves for South	Occasionally seen
Oystercatcher	Moulting	Many migrants arrive	—
Dotterel	Migrating	Migrating	Absent
Golden Plover	In large flocks	Frequent coast	Frequent coast
Gray Plover	Many arrive (chiefly young)	Still coming (adults)	,,
Lapwing	Quit uplands	Frequent coast	,,
Red-necked Phalarope	,,	,,	,,
Curlew	,,	,,	,,
Whimbrel	Still arriving on coast	Majority pass South	Few remain on coast
Common Sandpiper	Migrating	Migrating	Absent
Redshank	Many arrive from North	Frequent coast in flocks	On coast
Greenshank	Migrating	Migrating	Has occurred sparingly
Bar-tailed Goduit	Passes on migration	Passes on migration	Few remain for winter
Knot	Adults arrive	Continue to arrive	Frequent muddy coasts
Dunlin	Moulting	In large flocks on coast	,,
Purple Sandpiper	Arrives on coast	Continues to arrive	,,
Sanderling	Continues to arrive	Most have passed S.	Few occasionally seen

Species.	September.	October.	November.
Woodcock	Pairs separated	Migrants arrive	Continue to arrive
Snipe	Completes moult	Lives solitary	Still on moors
Jack Snipe . . .	—	Arrives	Lives solitary
Terns	Moving South	Moving South	Absent
Gulls	Spreading over seas	Follow shoals of fish	Roaming about
Skuas	,,	,,	,,
Puffin	,,	,,	,,
Auks	,,	,,	,,
Divers	Go out to sea	Roaming about	Roaming about
Grebes	Draw South and to coasts	Many passing along coast	Frequent fresh waters near sea as well as sea itself
Petrels	Rarely seen near land	Keep out to sea	Keep out to sea
Wild Swans . . .	—	—	Occasionally seen
Bewick's Swan . .	—	—	,,
Bean Goose . . .	—	—	Frequent low coasts
Pink-footed Goose .	—	—	,,
Gray-lag Goose . .	Wander down coasts	Migrants arrive	,,
White-fronted Goose	—	—	,,
Brent Goose . . .	—	—	,,
Bernacle Goose . .	—	—	,,
Pintail	—	—	More or less common on coast
Wigeon	Completing moult	—	Inland as well as coast

Species.	September.	October.	November.
Teal	Completing moult	—	Inland as well as coast
Shoveller	,,	—	,,
Mallard	,,	—	,,
Pochard	,,	—	,,
Scaup	,,	Arrives on coast	,,
Tufted Duck	,,	—	,,
Golden Eye	,,	Arrives on coast	,,
Long-tailed Duck	,,	—	More or less common on coast
Scoter	,,	—	,,
Velvet Scoter	,,	—	,,
Eider Duck	Moulting	Few wander South	Few wander South

WINTER.

Part IV.—Winter.

CHAPTER I.

THE TERRORS OF THE WINTER.

'Tis done ! dread Winter spreads his latest glooms,
And reigns tremendous o'er the conquered year.
How dead the vegetable kingdom lies !
How dumb the tuneful !   Horror wide extends
His desolate domain.

THROUGH the gilded autumn, animate and inanimate nature slowly and peacefully sinks into the lethargy and stupor of winter. Gradually the autumn days become shorter and more boisterous; the night frosts gild the herbage with a silver sheen ; the weather becomes more chilly, and the winds commence to sough mournfully through the woods, bringing down the last remaining leaves in fitful showers. With the earliest signs of winter's advent, even the most venturesome and daring of the summer migrants hurry away to warmer climes, and the fields and woods begin to have a very deserted and cheerless aspect. All the glorious autumn foliage lies in damp, decaying heaps on the sodden ground, the tall weeds stand brown and sapless, the noble ferns are prostrate, the bracken is fast crumbling away. The bare branches bend and lash like whips before the heightening gale, and even the very densest of thickets and coppices scarcely impede our vision now. The brilliant berries of the autumn still hang upon the briars and thorns, now mellowed and made more palatable by the early frosts, a winter garner for a feathered army. As soon as the deciduous trees and shrubs have lost their leafy covering, the evergreens stand boldly out in bright relief, the scarlet holly berries, and the wax-like seed of the mistletoe, which grows in yellow clusters on the poplars and the hawthorns, forming a garnish beautiful in the extreme. All these are signs of a changing season. Then the winter days creep on apace. Morning after morning the white frosts increase in magic beauty, and now and then a fitful shower of snow or sleet drives across the hills, an omen of keener weather yet to come.

Instinctively the various wild creatures feel the terrors of winter coming on; the birds crowd into the sheltered districts, or wander up and down the country unsettled and anxious; the animals that lie dormant through the cold hurry into their snug retreats, and in torpor await the spring; those not so constituted don their warm winter coats and bid defiance to the elements. Then comes the first heavy fall of snow, silently and suddenly in the night, changing the entire aspect of the country side. Winter in downright earnest is here at last. What a grand transformation a few hours' fall of snow will work! Overnight we gazed upon the green meadows, the dark brown fallows, and the leaf-scattered woods, much as autumn left them — a world bestrewn with the ruins of its summer grandeur; now in the morning all is changed into one vast canopy of white, gleaming and glittering in the sunshine! Overhead the bright blue sky seems eloquent enough of peace; the storm-clouds have vanished, but the air is sharp, and the wintry breeze rattles keenly through the naked trees, dislodging the big white masses of snow, which scatter and powder into crystal dust as they fall. What pen can do justice to the grandeur of a snowstorm—what words can properly describe this fairy scene of enchantment which Nature has so silently formed? The tree trunks are garnished with fleecy touches here and there in the crevices of the bark, and on the weather side of every one masses of snow have gathered as the storm in blinding fury swept by. In places sheltered from the winds, every branch and twig is covered with the snow, and all the undergrowth is wreathed and matted together with wool-like cables—a perfect lattice-work of brilliant white. Here and there the snow has drifted into banks smooth as glass, amongst which the tall dead weeds tower grimly, all of them capped with a white massive crown. Every post and fence and stump has received its decorative touch—snow, pure and dazzling snow is everywhere. Lost and bewildered birds fly across the white wastes which yestereven were green meadows and pastures. Already they feel the pinch of hunger, and betake themselves to the berries on the hedges, or to any green oases they may see from their lofty path. The hedgerows in places are almost buried by the drift, the fields are deserted, and even the wariest of birds draw near to the farmhouses—hunger overcoming their inherent wildness and timidity. Bowed down to the earth with their weight of snow are many of the branches of the evergreens; snow in masses clings to the foliage of the ivy on the trunks and limbs of the forest giants. The rabbits scratch and scrape amongst the snow to uncover the herbage on which they feed; the nervous hares start hurriedly across the white fields and seek the turnips; scarcely a bird is heard to chirp—animated nature is panic-stricken by the sudden, unexpected storm. Should the snow lie any length of time, it is surprising how soon the various wild creatures make the best of things, and accommodate themselves as well as they can to the changed conditions of their life. Many there are that hurry away to districts which have escaped the storm, returning to their old haunts with the thaw; but many more remain and pick up a precarious living among the snow. Most wild creatures leave the higher ground and seek shelter in the valleys; birds especially flock to the shrubberies and stackyards, forming for the time being into one great commonwealth — rendered sociable by the one common misfortune.

Then comes the thaw. A week or a fortnight has the snow-shroud lain on the ground, but much of its pristine beauty has vanished, the impurities of the atmosphere and the droppings of wild creatures sullying and staining its dazzling surface. The wind has shaken down the wreaths from trees and underwood, and scattered some of the drifts; everywhere the once smooth and gleaming pall has become disfigured by the footprints of animals and by drops of water which have fallen from the branches here and there as the snow melted in the sun. Sometimes a sudden change of temperature takes place, and a ground thaw soon removes the snow; at other times, the rise brings rain, which quickly washes all away, and the green fields, brown fallows, and gray woods appear once more, looking very grimy and dirty by force of contrast with what we have been accustomed to see so long. Patches of snow linger here and there by the hedges, and in the corners of the fields in places shaded from the sun; but all eventually disappears, and the snow-storm, with its terrors and hardships, becomes a dismal dream of the past. The drains and ditches pour the snow-water off the land into the brooks and rivers, often flooding the low-lying meadows, and then the various species of wild fowl congregate upon them; once more the country is "open," and birds and animals can find their accustomed food.

But winter still reigns supreme in spite of this spell of mildness. Once more the terrors of the season return, this time in a long-continued frost, which hardens the ground into adamant, and locks up the lakes and pools in an icy grasp. After a few days of calm, open weather, almost spring-like in its balminess, the cold returns, usually at eventide, and during the night the frost begins. Its fairy fingers bedeck the trees and grass with almost as much entrancing beauty as the snow; each twig, and spray, and stump is coated over with a film of ice crystals, each leaf of the evergreens and the privet trees is bordered with a silver margin. Day after day, sometimes for weeks in succession, the frost continues, and birds and animals perish in thousands from hunger. Cold rarely, if ever, kills, so long as these wild creatures can obtain sufficient food; they are well clothed, and take good care to sleep in warm corners, but when the supply of sustenance begins to fail, mortality is high. There is something to me exceedingly pathetic about a starved, dead bird. True, we find but few of them, for all wild creatures, when stricken by death, seek to conceal themselves, and quietly perish hidden away and alone. Those that seek their food in the ground are the first to succumb; many retire south, many more relinquish their usual habits for a time, and pick up a living near the dwellings of man. A continued frost is far worse than the heaviest fall of snow; and I always welcome the break-up of the spell of hard weather, for I know what sad havoc it works in the ranks of my feathered favourites.

With the return of milder weather, the moping, weakly birds soon become merry, and active, and strong again. Their sadness is short-lived; Nature abhors continued sorrow, and it is surprising how quickly birds, and animals too, get into good condition again as soon as the season of adversity is past. The frost releases its iron grip of the fields and streams, the long icicles melt like magic, and the swamps relapse into their normal state of sponginess again. Then, perhaps, ensues a week or so of really mild, genial weather. Winter is not all terrors—all ice, and frost, and snow; there are brief intervals of spring-like balminess, when the air is mild and the sunshine warm. On these fitful days the Thrushes pipe a few notes, and an occasional Skylark soars upwards into the blue sky to try over its matin song, to keep itself in tune until the spring. Many other birds assume unwonted activity; wild animals are full of joy once more, those dormant ones that sleep the lightest often coming out into the sunshine, but returning to their warm winter nests at the renewal of the frost.

But the winter is not over; the snowstorms and the frosts, the pitiless showers of hail and sleet, and the biting winds return, changing the fitful smiles of joy on Nature's face into aspects of fear and terror. The songsters hush their music, birds flock back to the shelter of the shrubberies and farmyards, the animals and insects that hibernate remain safe at home, and all is desolation and ruin once more. And so the long, dreary, northern winter drags its weary course along, its monotony relieved by occasional days of brightness; now frost and snow, then genial sunshine, anon wind and icy tempest. As the days slowly, so very slowly, lengthen, the cold only increases in intensity; but we are cheered now by the few extra hours of daylight—that, indeed, is a welcome sign the lowest depths of the wintry sadness have been reached, and we may now expect the spring. The sun is on his way back to Cancer, and the long, fierce struggle with winter has begun. Winter yields his iron sway over our northern lands reluctantly, and the days of February and March, the last of winter and the earliest of spring, are full of the warring of the elements. The warmer sunshine tempts the leaves on the whitethorns and the elder trees to burst their sheaths, but the night frosts shrivel them remorselessly; the early birds essay to build their nests during a fitful period of calm, but the last gasps of dying winter stay their labours and hush their vernal song. There is a certain solemn grandeur, awe-inspiring and impressive, about the season of winter, which is intensified during its latter days, when gentle spring is striving for supremacy with the snow-crowned giant. Winter, too, is a period of rest. Much of Nature seems to sink to sleep through the tranquil autumn, and the terrible ordeal of winter is passed in a deathlike trance. But the frost and the snow are beneficial to the soil, and the cold braces all organic life for the heats and the sultriness of summer. Then comes the fitful awakening of spring—the resurrection of life, when the victory over winter is celebrated with revel and with song.

So that, with all its terrors, the winter is by no means devoid of interest; the various habits of birds and animals in adapting themselves to its rigours and its hardships, the beauties of a white slumbering world, the grandeur of the tempest and the storm, are an endless source of instruction and amusement to him who loves to study Nature out of doors.

CHAPTER II.

AMONG THE BIRDS IN WINTER.

We have watched the birds in spring-time, as the flowing tide of life revived, and all living things acquired new vigour under its gentle influences; we have studied bird life in its summer aspect, among sunny fields, in cool green woods, and by the restless sea; we have been among the birds in the year's decline, when the rapidly-retreating sun lit up the woods, all glowing in their brilliant if dying tints, and the waning season sent the migrants back to warmer climes; it is now our pleasant task to trace the various habits of birds in winter, among leafless trees, under lowering skies, and through frost and snow. Nothing is more interesting than to watch these ever-varying phases of bird life through the year. Ever there is something fresh to engage the observer's attention peculiar to the season he may chance to take his walks abroad—some old favourites have departed, others have arrived to take their place; some have just regained their powers of song; some are congregating, others dispersing. Change upon change in the birds' economy is ever occurring to his observation, at some seasons so rapidly that even a day scarce passes but his knowledge is increased by a variety of interesting facts. In winter he may watch the ways of birds as they lead a gipsy kind of life, ever wandering in search of food; in spring, the arrival of the vast army of migratory birds is a salient feature, love and courtship, song and war, being the order of the day; in summer, domestic arrangements are most birds' all-absorbing care; whilst in autumn, their loss of voice, moulting, gathering together, and migrations afford abundant scope for him who loves to study wild bird life in field or wood, or by the lonely shore. Now we will go out amongst bird life in the snow, to watch our feathered friends on naked branches, and to follow them along the frozen streams, and into the warm shrubberies, where the evergreens afford them shelter from the storm.

Nothing disorganises bird life so much as a long-continued frost; like a great army in full retreat, the various species fly before it, each struggling for food and life, and compelled to alter their habits with the unusual change in the weather. Before the storm arrives, the birds unerringly foretell its approach. Vast flocks of Skylarks may often be seen winging their way across the gray, lowering sky, retreating before the snow which covers their favourite stubbles and buries their food. Another sure "storm-warning" is a flock of Lapwings hastening from the marshes, now frost-bound and covered with snow, to more open pastures. In irregular order, these handsome birds steadily pursue their way across country, the mark of every wandering gunner, whose long-waited-for opportunity is the season of the poor birds' adversity. At such a time many a storm-driven bird is seen in localities least adapted to its habits. Oceanic birds are driven inland; moorland birds seek the valleys, and marsh birds the troubled streams. Shy birds become tame; wary birds show an amount of trustfulness quite at variance with their usual habits. Gulls wander far up the rivers; Cormorants and even Petrels visit inland sheets of water. At such a time the air is often full of bewildered birds, careering aimlessly about, flying steadily along in twos and threes, or in immense flocks composed of many species banded together by one common impulse.

Then comes the threatened snowstorm, and the country-side becomes one vast white expanse. The soft fleecy snow covers everything. It even clings to the clusters of bright red berries on the hawthorns, where the Fieldfares and Misselthrushes have to shake it off before they can get their morning meal. See the various Thrushes how they congregate upon the berry-bearing trees this morning; Redwings and Blackbirds coming to them, now that the ground is covered with snow, and they are prevented from searching for worms and grubs which are much more to their taste. Indeed, the Redwing only takes to berries as a last resource, its food consisting almost entirely of animal substances. This bird suffers much privation during a long-continued frost. I have known them so tame at such a time as to allow themselves to be taken by the hand, and every particle of fat has wasted from their little bodies. They now frequent the banks of the streams, manure heaps, and even doorsteps and window ledges where crumbs have been thrown out for the Sparrows and Robins. In the late winter months, when the berries are all gone, the Fieldfare will do the same.

Many birds, however, show little concern after a heavy fall of snow. The Finches are seldom troubled by severe weather, so long as the different kinds of seeds on which they live are not hidden by the snow, which is rarely the case in this country. They feed on the stubbles, along the hedgerow sides, and on commons where thistles, docks, wild mustard, and other weeds rear their tall stems far above the snow-wreath. At this season of the year many hard-billed birds, such as Greenfinches, Chaffinches, and Buntings congregate near farmhouses and in stackyards, picking up the scattered corn, or pulling out the ears of grain from the ricks. The woods are made lively with little companies of Titmice that industriously search every twig and bud and cranny for insects and larvæ, or pay a fleeting visit to the elder trees and ivy to regale themselves on the luscious berries. The Woodpeckers, again, are never affected by frost and snow. Their food is always to be found amongst the timber, and their merry notes resound in the woods as they hammer away at the trees. They hunt in a more systematic manner, beginning at the foot of the tree and working gradually upwards, then dropping down to the bottom of the next tree to begin again. The Jays and Magpies lend life and animation to the wintery landscape as they scream and chatter among the leafless trees, or keep up a noisy concert at nightfall in the pine woods and holly trees where they roost.

When the snow has gone, many kinds of birds congregate on the newly-manured stubbles which have been sown down with clover — Buntings, Larks, Rooks, and Starlings searching for seeds and worms. Very early in the winter the large companies of Goldcrests and Titmice break up into smaller parties and spread themselves about the country, leaving the birch coppices where they assembled on their arrival here in autumn. Flocks of Geese may often be seen crossing the inland districts during winter, probably journeying from one coast to the other. These parties generally assume the shape of a wedge, or letter V; and occasionally I have seen them formed into a perfect A, the leader being at the apex. We must also pay attention to the various migratory movements of birds during this

[margin: Goldcrests disband, 4th December.]

season. Some of the most remarkable of these migrations are made by the Song Thrush and the Blackbird. These are all the more interesting because they are, strictly speaking, confined to winter, and are as yet very imperfectly understood. By the beginning of December, almost every Song Thrush has departed, the Redwings having taken their place; and only an odd bird or so will be met with until the new year is well advanced. In addition to the Song Thrushes that breed in the British Islands, there are many birds of this species pass our country on their way south—all appearing to migrate in flocks of varying size, and during the night. In the case of the Blackbird, the movement, though not so universal, is undoubtedly made, and the greater number of these birds are absent from their usual haunts during two out of the three months of winter. Another winter migrant is the Kestrel. This pretty Hawk remains with us as long as mice and beetles are plentiful; then the majority depart, only an occasional bird being seen until the return of spring. Almost every bird found with us during the winter is more or less a migrant, changing its haunts according to the state of the weather. After a snowstorm the Larks and Buntings will sometimes disappear from their favourite haunts for weeks, only returning to them when the weather opens again. Sometimes the Rooks and Starlings will wander off to some distant district, where food chances to

Song Thrushes have left, 1st December.

be plentiful, remaining away for several days, then turning up again in full force. Redwings will desert their accustomed haunts during a long-continued frost, and Fieldfares and Missel-thrushes are constantly wandering up and down like homeless nomads. Even such homely species as the Robin and the Hedge Sparrow will suddenly become scarcer during a spell of cold weather; and the Wren is ever varying in numbers all the winter through. Very often a district will become unusually full of birds, as they have been driven out of others by local storms. I have known, for instance, vast flocks of Chaffinches and Buntings to swell the ranks of our resident species, remain a few days, then just as suddenly depart. All the winter through, these local migrations are in progress, and may be noted by any observer who keeps a careful watch over the birds of any one district. My note-books for the past twenty years are full of instances of these winter movements; birds coming and going with every important change in the weather.

If some birds leave us in the winter, there are others that pay us uncertain visits. One of the most handsome and interesting of these is the Bohemian Waxwing. This wandering, irregular winter guest visits us more or less sparingly every season from the Swedish forests; but in some winters it comes in immense flocks. The last great Waxwing season was in the winter of 1866–67. The habits of this bird while with us are very restless, and its food in winter appears to consist almost exclusively of berries. The Waxwing breeds in large colonies, but is very erratic in its choice of a locality, a fresh one being selected every year in some district where food chances to be abundant. The Crossbill and the Shore Lark are also our winter guests. The former bird is a resident in this country, and the unusual "rushes" that make their way here belong to the class of gipsy migrants, which only wander southwards when an exceptionally inclement winter forces them to do so. At this season Crossbills are very tame, and climb about the fir trees, aided by their bill and feet, more like Parrots than Finches, calling sweetly to each other all the time.

Although it is midwinter, the banks of the stream are a chosen haunt of bird life. Where the water boils and foams round the mossy boulders, we are sure to meet with our old friend the Dipper. He is happy enough as long as the stream is open, and keen must be the frost that will stay its rapid waters. He dives into the ice-cold stream, in quest of his insect food, just as heartily as in the spring and summer, his dense plumage being impervious to the wet and cold. The charming little Gray Wagtails are gone; the Summer Snipe is on the lagoons of Africa; but the Kingfisher, in his beautiful dress of blue and chestnut, haunts the lower reaches of the stream. He is much less fortunate than the Dipper, and often feels the pinch of hunger. No birds are affected so much by a long-continued frost as those that seek their food in slow-running waters or amongst marshy ground. The poor Kingfishers fare badly at such a time, and numbers of them are starved to death, whilst they have even been found frozen to the twigs on which they have chanced to rest. Sometimes this bird may be seen sitting above the frozen pool, the banks all draped with icicles, and set in a framework of frost and snow, watching the tiny fish and water-insects on which it feeds, but quite beyond its reach, owing to the film of ice that has covered the water in a single night. In the alder trees on the banks of the stream, little parties of Siskins are busy picking at the seeds, and here and there a wandering Heron flies hurriedly away. This bird will not stay by the stream if the frost continues.

If the snow is everywhere, bird life is almost as ubiquitous. In the small swamp, which for some unaccountable reason has escaped the finger of the frost, we may flush the fat little Jack Snipe from his warm corner amongst the dead grass tufts. Unerringly he returns to his favourite winter quarters year by year, so that each season we may find him on precisely the same square foot of ground. He spends his summer far away on the Arctic tundras, yet with a marvellous memory returns to old familiar quarters a thousand miles from his nesting-place, travelling to them in the night. His relation, the Woodcock, is of a more wandering disposition, and it is only by the greatest good fortune we are favoured with a glimpse of him as he darts in erratic course amongst the trees. It is very remarkable how solitary the Snipes and Woodcocks are in their habits. None of them are gregarious or even sociable. You may flush scores of Snipes from a single marsh, yet every one of them lives there by itself, and manifests no interest in the doings of its neighbours. This habit is all the more extraordinary when we remember that the Snipes are a little group of the important family of Charadriidæ — perhaps the most gregarious and sociable of birds.

Bird life on the shore is little changed by winter's advent. True, we miss the graceful Terns, sporting fairy-like above the summer sea; but their place is taken by countless other birds that make our coasts their winter quarters when their home in the Arctic regions is uninhabitable. Vast flocks of Ducks and Geese haunt the water, and countless hordes of Sandpipers, Curlews, and Plovers trip along the muddy and sandy shores, following the ebbing tide, and sleeping or preening their plumage during high water, waiting till their feeding grounds are again uncovered. Moorhens and Coots congregate on the salt water broads and the estuaries, leaving the inland pools as soon as the frost becomes severe. Grebes often mingle with them.

Many a time, on the clear starlight nights of midwinter, the Moorhen careers about the air, uttering its shrill cry at intervals. This note differs considerably from the usual *cr-rick*, and is only heard when the bird is in the air. These birds are very fond of frequenting woods in very cold weather, and may sometimes be seen hopping and climbing about the bushes with much adroitness, and even roosting in the evergreens. The Coot is the most frequently seen on the sea of the two species, and its numbers are largely increased during winter by birds driven south with the frost.

The wild moorlands, in summer so breezy and enticing, now look particularly dreary and desolate, especially if covered with snow. All the birds of summer have gone—not even a Meadow Pipit remains. But the Red Grouse haunts them still, and finds his food in places where the snow has drifted. If the storm is long-continued, he will seek the farms in the valleys and pick up the grain in the stackyards. In very severe weather he often burrows deep down into the snow, and sleeps securely at night below the surface, safe in his warm bed among the heather. The Ptarmigan, in flocks at this season, and in plumage white as the driven snow itself, comes lower down the hillsides from its usual haunts on the mountain-tops. Still this bird loves the snow, for his plumage is in harmony with it, and renders him safe from the marauding Eagles and Falcons which scour the hills in search of prey.

Returning to more rural scenes, we find a stroll along the hedgerows and through the shrubberies by no means devoid of interest. To a casual observer the hedges seem deserted, but ever and anon the low complaining note of the Hedge Sparrow will draw the attention to that sombre little bird, as he glides shadow-like through the branches. Our resident Hedge Sparrows are increased in numbers every autumn by migrants from the north, which reach this country by way of Heligoland. Another point of interest in the economy of this species is the manner of its moulting. I am by no means certain whether the bird moults once or twice in the year—a very important circumstance affecting the position of this species in its scientific classification. According to the views held by present day naturalists, the Hedge Sparrow is an aberrant member of the Parinæ or Tits. But should it be proved to have a spring moult as well as an autumn one, its affinity with the Warblers would be established. In its spotted young it resembles the Turdinæ or Thrushes. I have certainly observed this species in the moult during February (February 22nd, 1879), and call attention to the fact so that field naturalists may further investigate the matter. Now and then a noisy Blackbird starts up from the sunk fences and flies just above the ground to a more secluded haunt, chattering all the way; and these situations are also a favourite retreat of the garrulous and ever active little Wren. Here and there a gay Chaffinch or a Bunting sits on the topmost sprays, resting for a moment as they pass over the snow-clad fields. But they must be ever watchful and on the alert, as the bold Sparrow-hawk courses up and down the hedge-sides, and may bear them off in an instant. In the neglected weedy corners of the fields and along by the hedges, where the thistle is allowed to flourish by the slovenly farmer, we may often meet with a party of Goldfinches. Beautiful little birds they are, displaying their bright colours to perfection as they cling with fluttering wings to the prickly thistle heads, or flit along from stem to stem, scattering the downy seeds, and twittering to each other as they go. Bullfinches, usually in pairs, pass along the tangled hedges, being particularly fond of dock seeds; and their low, soft, piping call-note is heard from time to time as one bird gets separated from the other. Both these species are thorough wanderers, and often travel miles in a day, feeding as they go. Nearer home we shall not fail to see the ever welcome Robin, so neat and sprightly-looking in spite of snow and frost, disputing with the Sparrows for the scattered crumbs.

Another interesting bird of winter is the Brambling, all the way from Arctic forests, a refugee from the northern winter. This species lives in flocks during the winter, and often mixes with Chaffinches and Redwings. It seldom wanders far from a chosen district, returning yearly to its old haunts, and if not molested is very tame and confiding. It loves to feed on beech-mast, and frequents the shrubberies and newly-manured fields. Rooks and Starlings are also birds of a wintry landscape, generally to be found in flocks near dung-heaps, and on pastures and newly-ploughed fields. In hard weather these birds often suffer severely from hunger; but they usually retire to more open districts if the frost continues long. Another little bird often seen near dung-heaps at this season is the Meadow Pipit. The Tree Pipit is migratory, and leaves the fields in autumn, when the Meadow Pipit quits the moors and takes up his residence on them. Both are migratory, yet the one retires to a warmer land, the other only from the hills to the lower grounds. A few Wagtails, Common Buntings, and Stonechats are also seen about manure heaps in winter. The latter bird is another of those species that shift their ground with the change of season. In spring and summer it loves the upland gorse coverts, but deserts them to a bird in the late autumn. The stubbles, when free of snow, are full of birds. Flocks of Skylarks live upon them; Lesser Redpoles, Linnets, Wood Larks, and Tree Sparrows pay them repeated visits. In the plantations and shrubberies adjoining the fields we sometimes come across a winter gathering of Hawfinches. This bird is more or less gregarious during the cold season, but is not very sociably inclined towards other species. Broods and their parents generally live together till the spring, and sometimes these family parties gather into larger flocks, especially where beech-mast or yew berries chance to be plentiful. Few birds are more skulking in their movements than Hawfinches, and very often their clicking call-note is the only sign of their presence. They fly through the yew thickets with remarkable speed, or hurry along from one beech tree to another, only showing themselves at uncertain intervals.

The occasional spells of mild weather we experience during the winter are apt to make birds forget the snow and frost, with all their attendant terrors, and to tempt them into unwonted activity. Skylarks and Thrushes feel the influence of the warm sunshine, and give vent to their joy in bursts of song; birds that live in flocks scatter more over the open country; localities which were deserted when the frost came, suddenly become full of birds. The Wild Ducks hasten back to inland waters; Larks return to the stubbles; and the various small birds that kept close to our houses when the snow was lying deep, now rapidly desert us and return to their wonted haunts in the woods and fields. Some of the most engaging of our winter birds are the various species of Titmice. These hardy little creatures lend animation to the woods and groves, and never fail to interest us by their droll antics and rapid movements. All the British species of this group are resident with us, though their numbers are increased during autumn and winter by birds from more northern lands. The Blue Tit is the most familiar, and is a constant visitor to the trees and gardens near houses. The Marsh Tit is the least so; but the Coal Tit and the Great Tit, the largest member of the group, especially during hard weather, are very tame and confiding. The Crested Tit rarely leaves his Highland forests, or the beautiful Bearded Tit the fastnesses of the fens; and the Long-tailed Tit loves the woods and fields at all seasons, no cold or snow driving it from its favourite haunts among the trees and hedges. A bit of suet or a bone is a never-failing attraction for the Blue Tit, and its comical ways and grotesque attitudes as it perches on these dainty morsels are amusing in the extreme. These birds often pick at fallen apples in the orchard, and frequently visit any stray fruit that the gales have left upon the trees.

Many a pleasant hour may be spent watching the habits of birds in the early hours of a winter morning. In spite of the cold, the majority of species are early risers, and many are abroad before the sun has risen. Here is a table showing the time of awakening of some of our common birds on the 17th of February. The sun does not rise until 7.12 a.m. on the morning of this day. Weather fine and bright, but cold, cutting wind from N.W.

Time.	Species.	Remarks.
6.30 a.m.	Robin	Uttering a few sharp call-notes.
6.35	Song Thrush	Singing sweetly; counted two birds only at present.
6.38	Blackbird	One commenced uttering its loud call-notes, which were answered at once by almost every other Blackbird within hearing.
6.40	Robin	These birds are now singing sweetly.
6.45	Wren	Heard this bird, but have no doubt it is stirring as soon as the Robin.
6.45	Song Thrush	Now in full song; counted six singing together in one small field.
6.50	Wren	Very noisy, but no song yet.
6.55	Chaffinch	Flew out of yew bush, evidently half asleep, and sat on a whitethorn adjoining, beginning to preen its feathers. Uttered one or two call-notes.
7.0	Greenfinches	Now astir: all seed-eating species rise late.
7.5	Chaffinch	In full song. All birds now awake. The east is red with sunrise, and music fills the air.

All through the winter the Rooks have regularly visited their old breeding-places. In spite of fogs and storms, snow and frost, these birds have assembled at the old familiar trees—usually first thing in the morning, but sometimes in the middle of the day, and occasionally not till evening. Sometimes only a few birds come, at others a vast flock, many of them being strangers. As the winter days pass on, the birds begin to make a longer stay, and one or two pairs now and then perch on their nests, as if to claim the ownership. They will also visit the nest trees twice a day now, but never remain to roost in them until the first eggs are laid. Another bird that visits its old breeding-place every day during winter is the Starling. Regularly as morning comes, these birds may be observed in pairs sitting on roofs and chimney-stacks, or on the branches of trees near their nest-holes. The Magpie also visits its old nest from time to time during the winter; and this habit seems to be peculiar to most, if not all, life-paired species. Home ties seem strong in birds that mate for life, and marks of affection are exchanged between the sexes throughout the year. These birds love to sit side by side, to fly and feed in company, and their mutually understood place of meeting is the old nest. The Jackdaw, the Raven, many of the Raptores, the House Sparrow, and some Titmice, are other birds that pair for life, and more or less regularly visit their breeding-places throughout the winter. For ten years I have known the Starling and the Rook conform to this habit without an exception!

Rooks clean out old nests, 17th February.

The songs of birds in winter must not be overlooked. Music and love with most birds go together; gray skies and wintry landscapes are not associated with either. Most birds lose their song in the autumn moult, and never warble again until the following spring; but to this rule there are certain exceptions. In winter the Robin is the most prominent songster, his sweet and plaintive strains being heard in every wood and coppice. He sings throughout the short winter days, even into the twilight, when the dull red sun settles solemnly down behind the hills. The restless little Wren is another winter songster, his loud voice ringing cheerfully out from amongst the icicle-draped roots and branches, through which he loves to hop and sport, with tail held impudently erect. Another winter chorister is the handsome Missel-thrush, or "Stormcock," the largest of the British Thrushes. His notes are usually given forth from the topmost branches of the highest trees, and resemble those of the Song Thrush and the Blackbird, but possess a wild cadence peculiarly their own. Far up among the bending branches, often before daybreak, his rich, wild lay is heard; the blinding snowstorm seeming but to increase the beauty of his song, and to lend it an additional sweetness. The Starling warbles right through the winter, and the Song Thrush commences to do so as soon as he returns to his accustomed haunts. This latter bird is first heard to sing according to the state of the weather, its migrations extending over three weeks. As soon as they are back again, if the season is mild, they sing very freely, visiting the

Song Thrushes return, 19th January to 10th February.

same perching-places daily. Rain will not stop their joyous song, but snow and frost silences them directly. In well sheltered districts, the Hedge Sparrow also contributes his simple little song to the winter concert. A mild day generally sends a few Skylarks warbling heavenwards, and the sweet-voiced Blackbird is heard to sing occasionally during exceptionally mild weather. Undoubtedly the Robin, of all this feathered orchestra, is the most pertinacious singer. Severe weather generally silences the singing birds of winter, yet he is never hushed. Many other sounds also help to swell this winter concert of the woods and fields. These are the noisy twitterings and merry call-notes of birds that have lost their proper song with the turn of the leaf. What, for instance, more cheery in the short winter days than the lively chorus kept up by a flock of Redwings or Bramblings, like peals of little bells, on the tree-tops; the harsh chatter of the ever active Titmice; or even the loud caw of the Rook, as he flies leisurely home at eventide? Such simple cries are passed unnoticed in the plethora of spring-tide music, but are welcome now, when every sound serves to relieve the monotony of the silent woods, and to tell, in prophetic strains, of better songs to come. It will thus be seen that winter is by no means the silent, songless season we are apt to think it. Even in the darkest days there are a few musicians to relieve the dreary monotony of a northern winter; and during the fitful spells of milder weather quite a concert of bird melody may be heard in all sheltered districts. The following table will show the state of bird music during the three months of winter:

Name of Species.	Dec., 31 days. Sang on	Jan., 31 days. Sang on	Feb., 28 days. Sang on
1. Missel-thrush .	31 days	25 days	24 days
2. Song Thrush .	Absent	16 days	28 days
3. Blackbird . .	Absent	Silent	Very irregularly
4. Robin . . .	31 days	31 days	28 days
5. Wren	25 days	15 days	18 days
6. Hedge Sparrow	Very irregularly	Irregularly	12 days
7. Willow Warbler	Absent	Absent	Absent
8. Chiffchaff . .	Absent	Absent	Absent
9. Whitethroat. .	Absent	Absent	Absent
10. Blackcap. . .	Absent	Absent	Absent
11. Redstart . . .	Absent	Absent	Absent
12. Starling . . .	31 days	31 days	28 days
13. Meadow Pipit .	Silent	Silent	Silent
14. Tree Pipit . .	Absent	Absent	Absent
15. Chaffinch . .	Silent	Silent	18 days
16. Yellow Bunting	Silent	Silent	15 days
17. Greenfinch . .	Silent	Silent	Silent
18. Skylark . . .	Very irregularly	Irregularly	Last 10 days
19. Cuckoo . .	Absent	Absent	Absent

Before the winter quarter is past there are many little incidents occurring which serve to remind us of the coming, though yet far distant, spring. Some of the earliest-breeding birds pair during the latter part of winter, and many social gatherings disperse. By the end of January, House Sparrows begin to show signs of breeding, visiting their old nesting-places, and becoming very pugnacious and noisy. The earliest nests are almost invariably in holes of buildings, those in the trees not being used until later in the spring. I have known, in exceptionally mild winters, fresh eggs of the House Sparrow as early as the first week in January (6th January, 1877). By the middle of January, Hedge Sparrows are nearly all in pairs, and a month later the coveys of Partridges begin to disperse for the nesting season. The Chaffinches are now rapidly assuming their wedding finery, all the pale margins which have concealed the showy feathers rapidly abrading. About the same time, the Song Thrushes and the Blue Titmice pair, and the Great Tit begins to utter those notes which are peculiar to the season of courtship and love. By the end of the month the Blackbirds have chosen their mates, and, so far as birds are concerned, the vernal season has commenced. The Wood Lark is another of the very first birds to change its habits at the turn of the year. Having remained silent since the autumn, it begins its lovely song early in February, sometimes, in mild weather, during the last few days of January. Few birds are more locally distributed, and everywhere it prefers light, sandy districts to those which are heavy and wet. All through the

[margin notes: Few Sparrows begin nesting, 14t February. Partridges pair, 15th February. Chaffinches pair, 17th February.]

winter the Wood Lark is gregarious, and often mingles with the larger gatherings of Skylarks. When flushed, these birds often take refuge in a tree; they live principally on small seeds at this season.

Far out at sea there is much of interest to be seen among the birds. Many are the species gathered together where the shoals of fish are numerous. The rock-birds, such as Guillemots, Razorbills, Black Guillemots, and Puffins, wander long distances from the places where they breed, following the fishing fleets, and taking their share of the harvest of the sea. All the winter through these birds remain off our British coasts, leading a nomad life, until the grand reunion in the spring at the great breeding stations. At this season, too, the Great Northern Diver wanders southwards, and may often be seen frequenting estuaries and quiet bays. Many of these birds are only seen in our southern seas during winter. They live in the Arctic regions, or in the extreme northern portions of the kingdom, and go back again to their summer haunts at the earliest dawn of spring. Sometimes these birds fare badly during the wintry gales, and I have known the shore for long distances strewn with their bodies after unusually boisterous weather. At this season we have also a great many favourable opportunities for observing the habits of many species of Ducks which only visit this country during the winter. Both at sea and on the inland sheets of water, these fowl congregate in enormous multitudes. When they first arrive they are exceedingly tame and confiding, but incessant persecution soon teaches them wariness, and even in the severest weather they are careful not to allow too close a scrutiny. It should here be remarked that our well-known Mallard, as well as most other Ducks, pair during the winter. This is the season when these birds are in the gayest and most brilliant plumage—a time, by the way, when all male birds pay court to the opposite sex. Did they delay this matter until just previous to the breeding season, the wedding finery of all the northern species would have lost the greater part of its lustre. These Arctic Ducks do not breed until June, when the short northern summer is just about commencing.

It might be thought that bird life in winter displayed too little variety to tempt the observer out of doors; but no greater mistake could be made. There is a novelty about the habits of birds at this season which will not fail to impress the beholder with its charm, and to fill his walks abroad, during the months of frost and snow, with feelings of deepest interest.

CHAPTER III.

SOME VISITING CARDS.

There are many branches of woodcraft in which the observer must make himself proficient ere he can hope to become familiar with the ways and habits of birds. Tracking is one of them. Birds leave many signs of their presence behind them, one of the most important being their transient footprints on the shining sands, smooth soft mud-flats, or white expanse of snow. To the patient observer who takes care to see everything and to make himself familiar with its meaning, these tracks are so many visiting cards, by means of which he is able to read the names of the birds that left them behind, and to learn much of their movements. These footprint hieroglyphics have a story to tell—observation is the key by means of which they may be read. Before attempting to decipher any of these footprints in the mud or snow, the observer must pay considerable attention to the formation and general characteristics of the feet of birds. These vary to such a remarkable degree that in many instances the particular species to which they belong is easily determined by this character alone. Having thus made himself acquainted with the structural peculiarities of the feet of birds (not from dried skins, by the way, but from birds in the "flesh"), he must then become familiar with the manner in which these feet are used. Some birds take long and stately strides, others take short steps; many species progress in hops, both feet held exactly side by side; whilst others, yet again, run quickly from place to place. Again, there are some birds with tarsi so short that they leave impressions of their bodies behind them as well as footprints.

Having acquired this preliminary knowledge, let us go out on to the mud-flats on some wintry morning. The inland woods and fields and roads are lying deep in snow, but the muds are brown and dismal-looking, their dreariness emphasized by the strong contrast of colour. The tide has long ebbed, leaving us a clean slate, as it were, for the curious writing we have come to find and read. Here and there on every side are the runes of the feathered race that haunt this vast expanse—footprints straying up and down, tracks crossing and recrossing in a perfect maze of confusion. Footprints are everywhere—round the margins of the little tide-pools, imprinted in the soft mud at the bottom of many of them, as the wading birds that made them tripped through the shallow water; up and down the broad smooth spaces, on the tiny mud-hills, in the hollows, and by the very edge of the receding sea. The birds of the sands and mud-flats are shy and wary enough in winter; not one is to be seen within a quarter of a mile of where we are standing; but they have left their "cards" behind them on the shore, and we can read from them the story of their lives. Here the mud for a space of many square yards has been trampled into a rough, uneven surface. What birds are these, which in crowded assemblage have left their marks behind them? On the edge of this mass of footprints there are several impressions clear and sharply defined as when the birds stamped them on the muds. They are broad and webbed, the tell-tale mark of Geese; and see, some of them are much smaller than others, a sure indication that the flock was a mixed one and composed of several species. A few feathers left sticking in the mud and floating on the neighbouring tide-pool show that Brent Geese have been here. An assembly of Geese has been sleeping here, waiting on the higher ground for the tide to ebb. They have now gone off to feed—even as we are examining their footprints the familiar *gag, gag, gag,* sounds faintly from afar, where the noisy host are making their morning meal. We notice more webbed footprints farther on; larger ones still this time, and undoubtedly made by Swans. These impressions are also of different sizes, the larger ones being six inches or more from heel to toe, the smaller ones an inch less. By our knowledge of the feet of birds, we are able to identify the large ones as those made by the Hooper or Wild Swan, whilst those of the latter were left behind by the much smaller-footed Bewick's Swan.

All along the margin of this shallow backwater at regular intervals big footprints are stamped upon the mud; they are quite seven inches from the tip of the middle claw to the end of the hind one, and are webbed at the base of the third and fourth toes. A Heron has been here some time during the early hours of morning. Not only has he wandered round the mud on the shore, but he has waded into the pool, for there are his footprints at the bottom of the water; and see, he has stood here for some time on his left leg, evidently in moody contemplation, or perhaps asleep. Here on the bank of the stream, just where it flows into the sea, are one or two clear impressions, which undoubtedly were made by a Cormorant. How are we thus able to speak so confidently, for the bird that made them is most probably now far out at sea? Simply because these footprints are those of a web-footed bird, and all the toes are joined together by a membrane. Now, the Cormorants are the only group of known birds in which the hind toe is connected by a web with the other toes, so that the footprints can never be mistaken. The Shag has a smaller foot than the Cormorant, and that of the Gannet is larger. Across the open muds the Curlews have been running; here are their large footprints mingled with those of smaller fowl; and there are the singular impressions left by the lobed feet of the Coot. Here and there are the short thick steps of the Oystercatcher on the mud banks, conspicuous by having no imprint of a hind toe, which is wanting in this species. These pretty birds have been digging for cockles in the sand; we heard their loud wild pipe of alarm a quarter of an hour ago, as they rose into the air and sped along the coast to quieter haunts; the writing they have left upon the mud tells us plainly what they were about.

A little further on, the mud bears the marks of many feet, imprinted there by wanderers up and down the shore. There, by the rippling waves, the Sanderling has been tripping to and fro. He feeds close by the wash of the tide, and we may know his tiny feet from all the rest because the hind toe is wanting. Other birds have also been feeding here and left their tracks behind them. These are the impressions made by the members of the genus Tringa, identified by having all the toes cleft to the base. A flock of Knots have rested there, and spread out in regular order, all with head to wind, to search systematically every foot of ground. This flock took their departure suddenly and in a compact mass, for not a stray outlying footprint is anywhere to be seen. The Dunlin, with his more delicately formed foot, has also been feeding close by; three hundred yards ahead of us, the flock is still upon the shore, a restless, busy company. Already they are alarmed at our presence, and, moved by a common impulse, a thousand wings are spread together, and the wary birds are gone far down the coast, their pinions gleaming like silver in the sun. Easily distinguished also are the footprints of the birds forming the genus Totanus, in all of which the outer and middle toes are united by a membrane at the base. The dainty Redshank leaves the erratic tracery of its orange-coloured feet upon the mud more frequently than any other birds of this group; but sometimes the larger footprints of the Greenshank and the Goduits may be seen. Many Gulls have rested on the mud from time to time. Those impressions at our feet have left a tale behind them. We can picture how a large Gull swept gracefully down from the air and alighted here. There are the two footprints side by side, made by the bird when first dropping on the mud; then we can trace the walking to and fro; and finally, the last few footsteps as this Gull ran forward, expanding its long wings, and then rising fluttering into the air. All this is plainly written here, as clearly and distinctly as though we absolutely saw the bird in the act of inditing it!

Now let us leave the mud-flats, and wend our way inland to the distant fields and woods. We shall find writing there as interesting, though the tale it tells will be a different one. How smooth and white the country looks; snow tones down all the rugged roughness of the roads and hedges, and softens the outlines of the hills. Beautiful are the woodlands now, clad in foliage white and fair; blossoms once more deck the dead hemlocks and the bare hawthorns—blooms of brilliant snow. The hedges are garlanded with fleecy masses; the long brambles, and briars, and withered bines of the honeysuckle have all been transformed into nets and ropes of snow! Nothing yet has sullied the surface; no carts have lumbered over the tracks by the hedge-sides; the ploughshare is half buried at the end of the furrow where it was left when the frost came; the cattle, sheep, and horses are safe sheltered from the storm at home, down yonder at the farmstead in the valley; not even the keeper's tracks across the fields and along the drives in the white woods have yet disturbed the glittering canopy of snow. A Robin calling in the laurels, and a Wren reeling off its noisy song amongst the brambles, are almost the only sounds that break the stillness of this wintry death scene. Nothing seems stirring; yet there is abundant evidence written on the snow-wreath that many creatures are abroad this morning.

Here, out on the open fields, a hare has passed leisurely along; the tracks show that "puss" was in no hurry, though evidently bewildered, for her spoor is up and down in a very undecided sort of manner. Along the wood-side the rabbits have been romping to and fro through the snow-decked hedges, and back into the copse; there a weasel's tracks are printed from the hedge all down by the side of the stream, in and out of the thickets of rose briars and thorns, under the snowy arches of bracken leaves, and finally lost in the hedge again. He has been taking a morning stroll in quest of breakfast; but rabbits are too wary, and birds not numbed enough yet for an easy capture. His tracks are not marked by blood, as is so often the case; he has gone home hungry as he came. As we enter the wood, another style of writing may be seen upon the snow. Up and down the drives, and round the places where food is scattered by the keeper, the three-pronged feet of the Pheasant prick deeply into the surface; his hind toe leaves little track behind, because it is placed some distance above the plane of the rest. Other birds have been down upon the snow as well as he. There is the short, quick stride of the Ring Dove, the surface of the snow being brushed away by the body of the bird as it walked, owing to the extreme shortness of the tarsi; there are the two footprints, side by side, of the hopping Jay, and mingled amongst them are the tiny impressions left by some small Passerine bird—most probably a Robin, as it dropped down from the bushes in quest of food. On the borders of the wood, down by the holly trees, the footmarks of the Blackbird are plainly visible on the snow. He fluttered out of the evergreens at dawn, scattering the loose snow from the prickly branches, and hopped along by the laurel bushes; then the track is lost, no snow lying under the trees to receive the impress of his feet. But see, there the steps begin again on the other side—a series of hops; then suddenly they cease. Our Blackbird rose here with startled cry, and made for the thorn bushes yonder. He is even there now, feeding on the scarlet berries!

High on the bleak stubbles, where the snow lies like fine powder, the running footprints of Skylarks may be traced, the long hind claw being perfectly visible in many of them; whilst in the pastures the well-defined and somewhat clumsy steps of the Rooks are everywhere, and mingled amongst them the smaller footprints of the Starling. Both these birds walk to and fro, and consequently their imprints are never side by side. Here are some in pairs, though, by the margin of the pond, made by the Redwings; and the hardy Wagtails have also been and left their cards upon the snow. Another track is visible here by the water—a big, bold footprint this, four inches or more from toe to toe; it has been made by the long, slender, pliable feet of the Water-hen. We can trace it some distance out on to the open field, then back again to the water-side, along the rush-grown margin to the thicket of yellow iris leaves, in which the bird itself is skulking now. The curious lobed feet of the Little Grebe have also left their mark behind them. Rarely does this species come upon the land, and its footprints are not often seen in the snow; but those of the Coot frequently tell the tale of their owner's wanderings ashore. By the side of the swamps the long, slender footprints of the Snipe, each one cleft to the base, may sometimes be seen; and in the game coverts by the stream the Woodcock leaves the unmistakable sign of its visits stamped upon the snow.

Many birds there are that flit among the trees and bushes, yet never leave a record of their presence on the snow. Such birds as Titmice rarely or never come to the ground during a snow-fall, and Nuthatches and Woodpeckers invariably keep to the tree trunks and branches. We must read their various signatures in the dislodged snow from the trees; but at best the writing is blurred and indistinct, and difficult to decipher. Sometimes a snow-covered branch will display the footprints of a bird that may have chanced to perch upon it, but the discovery of such is rarely made. Here, for instance, a bird has been sitting on this snow-crusted branch hanging over the stream. The footprint left thereon is no ordinary one. It is a tiny stamp, indeed, and the two outermost toes in front are joined together for nearly half their length; the mark of the Kingfisher. On rare occasions the Woodpeckers leave their tracks upon those trunks which the driving wind has decked in a snow-wreath. They are readily distinguished—two toes in front, two behind; but the Woodpeckers shun the snow, for it fills up the crannies of the bark and prevents them from finding the various forms of insect life that lurk therein.

In the farmyard, many little footprints are traced upon the snow—left there by Sparrows, Buntings, and Chaffinches; whilst in the hedgerows an occasional "card" is left by the skulking Hedge Sparrow and the restless Wren. By carefully following up and studying these tracks upon the snow, we are enabled to read the various actions of the birds at the time they imprinted them there. We can see how the Rooks have been walking up and down, digging here and there into the ground for some lurking grub or other food; we can tell that the Starlings visited these excavations after the Rooks had left them, in the doubtful hope of meeting with something suited to their own particular taste. We can track the Robin's path from the wood-heap to the scattered crumbs outside the cotter's door; we know the errand this bright-eyed songster was on; we can picture him in imagination taking every one of those tiny hops, flicking his wings and tail the while, and looking so trim and neat as he came along so full of confidence. In fact, these writings on the mud and the snow are full of birds' habits; they are a faithful record of the ways and doings of our feathered friends, if we only take the trouble to read them.

There are other tracks upon the snow which it may be well just to mention, for they reveal the presence of wild creatures, and tell us much of their habits and economy. Here the brilliant purity of the snow-shroud is stained and sullied with drops of blood—murder has been done this morning among the dark foliage of the pine trees. This is the mark left by the bloodthirsty Sparrow-hawk; and a little farther on a heap of feathers completes the tale of death—a Chaffinch, driven to the firs for shelter from the snowstorm, has met his doom. The same wood has been the scene of another tragedy. This time a rabbit has raced for his life through the snow-wreathed brambles and under the broken fern. A few yards farther on his blood stains the snow; a weasel has ridden him to death; with sharp teeth meeting in his neck, he has sucked his life's blood, and there under the laurels he lies cold, and stiff, and dead!

Tracks of birds are also traced upon the snow on the moors. Birds up here are scarce in winter time; still the footprints of the Red Grouse, broad and bold among the heather, show distinctly; whilst the mountain-tops are traced with the moccasined feet of the Ptarmigan; and in among the pine woods, lower down the valley, the big footprints of the Capercailzie and the smaller ones of the Black Grouse are frequent enough amongst the trees and on the broad branches.

Such is a little of the interest to be derived from tracking — from tracing out the various hieroglyphics left by wild creatures on the muddy shore and in the snow. In following these footprints of birds and beasts across the white wastes, we cannot help materially increasing our knowledge of Natural History. It also serves to sharpen our powers of observation, and to render us more acute and cunning in circumventing these wild and wary creatures of the wilderness. The charm of a walk abroad in quest of information after the snowstorm, either in the woods and fields, or across the vast expanse of sand and mud on a coast where birds are plentiful, is only known to those who have taken it in full sympathy with Nature and her works. To a lover of birds this is an exceptional treat, for many a pleasant chapter of Ornithology is written on these brown scrolls of mud and shining sheets of snow!

CHAPTER IV.

SOME BIRDS OF THE WINTER.

THERE are one or two birds of special interest to the naturalist which may very fairly be classed as birds of the winter. Some of them may make their appearance on our shores during the autumn, but it is with winter that they are most closely associated. All of them are birds of the Arctic regions, and all belong to that class of feathered travellers to which the name of gipsy migrant has been very appropriately applied. They are birds which have no regular winter home. Like the nomad tribes of the wilderness, they wander to and fro, south and north, just as the exigency of the weather drives them. So long as they can pick up a precarious sustenance among the snow, on the beaten post-roads and round the northern villages, the majority of these birds are content to pass the winter in the higher latitudes. Comparatively few of them extend their wanderings far beyond the limits of the severe weather; and of those that do, a certain number of most of these species reach our islands every winter. Sometimes a more rigorous winter than usual drives these gipsies southwards in greater numbers, and then they pay their uncertain visits to us in flocks or "rushes." This is especially the case with the Waxwing and the Shore Lark. The habits and geographical distribution of these northern strangers well merit a special chapter in the winter annals of our British birds.

First upon our list is the Waxwing. This remarkably handsome bird is allied on the one hand to the Starlings, and on the other to the Shrikes; most probably its affinities are nearest to the latter birds. The geographical distribution of the Waxwing is as yet but imperfectly known. There is, however, little or no evidence to show that the range of this bird is circumpolar; it is one of the few Old World species that have extended their habitat into North America by way of Behring Strait. The breeding range of the Waxwing is practically confined to the region of pine woods near the Arctic circle in Europe and Asia. As before stated, winter home this bird has none. Its wanderings in the cold season extend southwards into Central Europe, Turkestan, South Siberia, Mongolia, China, and Japan. The well-known but much smaller Cedar Bird of America is the Waxwing's nearest ally and its New World representative; but in Japan it is replaced by a beautiful species, remarkable in having the tail tipped with red instead of yellow, and the wax-like appendages on the shaft tips of the secondaries are replaced by red spots on the feathers.

In its restless habits the Waxwing very closely resembles the Titmice; it is also just as gregarious, and wanders up and down the country in parties and flocks. During their erratic visits to us, these birds seem specially fond of the hips of the wild rose, elder-berries, and the fruit of the service tree. I once followed a party of these charming birds for a considerable distance through the fields and spinneys. The snow lay thickly on the ground, and very handsome they looked, in contrast with the brilliant white wreath that decked the trees. From thicket to thicket, and from tree to tree, they flew in irregular order, climbing about the branches like Tits or Nuthatches, and turning and twisting into many different attitudes. They were remarkably tame and confiding, apparently not having then learned the lesson of wariness which relentless persecution soon teaches most of our distinguished feathered strangers. I often wonder how many of these various interesting visitors ever get back home again, or even survive the first few days of their stay! I have also remarked that the Waxwing is much more wary when alone than when in company with others of its kindred. The flight of this beautiful bird is rapid and well sustained, its wings being well adapted for long journeys. During its sojourn in this country it very rarely visits the ground, obtaining most of its food in the branches

Perhaps the most interesting part of the Waxwing's economy is its habits during the season of reproduction. For many years the nest and eggs of the Waxwing were unknown to science. As is usual in such cases, the cabinet scientists invented a tale of their own; and the belief that this bird reared its young in hollow trees and holes in rocks, in unknown forests, universally prevailed. The halo of romance surrounding the nidification of the Waxwing was not cleared away until the pluck and enthusiastic perseverance of Wolley were rewarded, in 1856, by the discovery of its nest and eggs. Even then Wolley had not the rare pleasure of finding these treasures for himself; they were obtained by his servant. Since Wolley's discovery, all the secrets of the Waxwing's habits during the breeding season have been made known; and certainly the manner of this bird's reproduction is surrounded with no ordinary degree of interest. Waxwings, curiously enough, breed in societies, like many other northern birds, such as Fieldfares, Redwings, and Bramblings; and, what is more interesting, the same locality is rarely used twice in succession, the birds apparently selecting districts where food is unusually abundant, as is the way of the Rose-coloured Pastor. The nest is placed neither in holes of trees nor rocks, but in the branches of the spruce and other fir trees; neither is it domed, but open, like a Shrike's, made of twigs, moss, and lichens, and lined with feathers. The eggs are from five to seven in number, various shades of green and gray in ground colour, spotted and blotched with light and dark brown and violet-gray. The young of the Waxwing are nothing near so beautiful as their parents, in this respect being like Starlings and Shrikes, and are streaked on the under parts. We never see them in this plumage of youth in our country, for they moult their dull dress in autumn, before beginning their nomad life.

The second species which we have to notice is the Shore Lark. This bird has a very wide range, being found in the Arctic regions of both hemispheres. It is a bird of the open tundras, beyond the Arctic circle, and only breeds above the limit of the growth of trees. Shore Larks also have no place of winter retreat, and spend that season battling with the frost and snow, coming south in severe weather, going north with the return of a milder period. At this season the Shore Lark gets as far south as Central Europe, Turkestan, Southern Siberia, and the north of China, and in the New World into the Southern States. It is remarkable that such an Arctic species as the Shore Lark has several very near allies resident in the south, even in as low a zone as the Sahara in the Old World, and Mexico and Central America in the New. These southern Shore Larks are probably the descendants of a common ancestor, driven southwards by the glacial ice — settlers which remained behind when the ancestral Shore Lark went north again to its old home round the Pole. It is worthy of remark that the Shore Lark is now much more frequently observed on the British coasts than formerly. Either this is because rare birds are more carefully looked for, and the number of observers is greater, or the bird is gradually assuming a more regular winter migration. In the absence of sufficient data it is, however, hard to decide.

When in this country, the Shore Lark is rarely seen far away from the coast. Strange as it may seem, the habits of this bird are better known during the summer, when it is in the far north, than in the winter, when it is more or less common in civilised countries. It is gregarious and sociable in its habits, mixing repeatedly with Lapland and Snow Buntings, as well as uniting into flocks of its own kind to wander south in winter, and return north in spring to its breeding grounds. The Shore Lark, strictly speaking, is a ground bird—of all ground birds it is one of the most terrestrial, its usual mode of progression being either a run or a walk, although it is capable of hopping if occasion requires. During its stay with us, it picks up its food upon the beach and the wild rough land near the sea; this is principally composed of various small seeds and buds, but in summer its diet is more insectivorous.

During the love season, the male Shore Lark

soars high up in the air to warble his simple song, rising and falling like the Skylark, singing as he goes. The nest is made in very similar situations to those selected by our own Larks—hollows in the ground—and is composed of dry grass, scraps of moss, and frequently lined with hair. The four or five eggs are remarkably like those of the Skylark in colour, and are about the same size. The general similarity of the eggs of all the Larks is a noteworthy fact; and all of them are of those olive tints which are least conspicuous in the situations where the nests are made. This family likeness also extends, in a great measure, to the young birds, which are more or less uniformly spotted in their first or nestling plumage. Young Shore Larks are very prettily spotted with yellow, something like the downy stage of the Golden Plover, which serves admirably to conceal them from danger on the many-coloured tundra where their infancy is passed.

Our third little bird of winter is the Pine Grosbeak. Rare and irregular are the visits of this Arctic species; yet it is possible that it may occur here more frequently and be overlooked. Like the preceding species, the Pine Grosbeak is a circumpolar bird, but instead of living on the tundras beyond forest growth, its home is in the pine woods near the Arctic circle. Of all the gipsy migrants this bird appears to wander least from its native woods; this is probably because it obtains its food from the trees, whereas the others seek their sustenance on the ground, where the snow often buries the seeds. It appears to become restless, however, in winter, and occasionally wanders to the northern districts of Central Europe, to Southern Siberia, Kamtschatka, and the Northern States of America.

Although the bill of the Pine Grosbeak is not absolutely crossed, as in the Crossbills, the upper mandible spreads considerably over the lower one; whilst the style of colouration, the various changes of the plumage, as well as the bird's general habits, all show its affinities with the Crossbills. In many of its habits it reminds us of the Hawfinch. It loves to frequent the same kind of country—open districts, where the trees are scattered up and down in picturesque groves rather than collected into forests, and is shy and retiring, flitting quickly in an undulating, Finch-like way from one tree-top to another, concealing its showy dress as much as possible amongst the foliage. Sometimes it may be noticed quietly sitting on the topmost spike of a spruce fir. The food of this bird is largely composed of fir-cones, berries, and the buds of trees; but in summer this diet may be varied with insects, as is the case with the other Finches.

The short but sweet love-song of the Pine Grosbeak begins in spring, and when two or three birds are warbling in the same grove, the effect is very pleasing and beautiful. The nest is generally made in the branches of the spruce firs, not among the fine sprays, but usually on a thick branch close to the tapering trunk. It resembles that of the Hawfinch in the method of its construction—the outside formed of fine twigs, the inside of roots, mosses, and grass, lined with hair. The eggs are four or five in number, very similar to those of the Bullfinch, but are nearly twice the size. The eggs of the Pine Grosbeak were another of the prizes brought home by Wolley in the year previous to his discovery of the eggs of the Waxwing. No eggs of this bird were known to science until these were obtained in the forests of Russian Lapland.

The last of our winter birds claiming special notice is the charming Snow Bunting. No other Passerine bird is more thoroughly identified with the Arctic regions than this little northern stranger; no other known bird penetrates farther into the Polar world, for its cheery chirp and still more cheery song enliven the wildest country in the highest latitude yet reached by man. The Snow Bunting is a circumpolar bird, its range extending across the northern portions of both the Old and New Worlds. It breeds on the Arctic tundras, far beyond the zone of trees, and on the high mountains further south, even as low as the Grampians, where the elevation ensures it a similar climate. During winter it wanders southwards into Central Europe, the southern districts of Siberia, the north of China,

Japan, and the most northerly of the United States. It is a noteworthy fact that none of these four winter birds whose habits we have been briefly studying, have ever been observed in the Spanish Peninsula.

The Snow Bunting is pre-eminently a bird of the snow; and its arrival on our shores, more likely than not, is the first warning of a coming snowstorm. I have watched these pretty birds in flocks among the wild, weedy grounds below the sea-banks on the Lincolnshire coast—the same county, by the way, where Willughby obtained it more than a hundred years ago, and thus established its claim to rank as a British bird; although there can be little doubt that the Snow Bunting regularly visited our islands since glacial times, and shared the Lincolnshire marshes with birds and beasts whose race has long been run. They spend by far the greater portion of their time upon the ground, searching for small seeds and buds among the rough herbage; but if trees are near, they will frequently alight upon them. Snow Buntings are also birds of the coast, most at home on the wildest beaches, in those districts most resembling their Arctic haunts; but sometimes a severe storm will drive them inland, and then I see them feeding on the stubbles and clover fields, in company with Bramblings and other Finches. Upon the ground this pretty bird both walks and hops; and its flight is more straightforward than that of its congeners, although the wings are closed every now and then, as is the case with them. During the whole time of its sojourn in this country, the Snow Bunting lives almost entirely on seeds of various kinds, which it seeks among the weeds and herbage; but in summer it appears to be almost entirely insectivorous.

In many of its habits the Snow Bunting resembles the Shore Lark—it frequents much the same kind of country in the high north, and, like that bird, it soars into the air, and warbles its short but sweet and charming love-song. The Snow Bunting does not appear to separate into pairs until it arrives at its breeding grounds. In flocks it passes northwards in the spring, keeping on the margin of the snow-wreath, pushing on as the signs of returning summer spread around; those birds that go the farthest north not reaching their old haunts before June. As the breeding season comes on, the Snow Bunting's beauty increases—all the pale margins of the feathers drop off or abrade away, until the plumage is pure black and white. The breeding season commences as soon as the birds can get back to their old haunts; the Arctic summer is short, and no time must be lost. The nest is placed in a variety of situations. Sometimes it is built among heaps of stones, in a similar place to that chosen so often by the Wheatear, or in crevices of the rocks; sometimes in heaps of drift-wood, and other *débris*, lying scattered on the shores of the Arctic Ocean.

There is a rustic charm about the nest of the Snow Bunting which makes it an exceedingly pretty structure. Outwardly it is formed of slender twigs and roots, then moss and grass, and finally a lining of hair, feathers, and occasionally vegetable down is added. The eggs—not at all typical of a Bunting, by the way—are six or seven in number, and bluish or yellowish-white, spotted and blotched with rich brown, and sometimes streaked with darker brown. Only one brood is reared; there is no time for a second, although the birds will lay again should their first eggs chance to be destroyed; and, as soon as the young can fly, the old roving life is resumed, and our restless Snow Buntings become the veriest of nomads once more.

The fitful, uncertain wanderings of these winter nomads illustrate very forcibly the attachment which birds have for their true home—an affection which is deeply rooted in the uncounted ages of the past. Most northern birds undertake a regular migration in autumn, because food fails them in their accustomed haunts; others only wander southwards just as far as they are compelled by the weather; yet all unerringly return home as soon as that home country becomes habitable. There is nothing, so far as we can learn, to prevent such birds as Redwings, Fieldfares, and Bramblings remaining behind to breed with us in England; yet their love for home overcomes every other inclination, and back to the Swedish woods and fells they go with the spring. I am strongly of opinion that these birds require a certain temperature in which to rear their young, and return each season to districts where such prevails. The Snow Bunting is a remarkable instance of this. In Scotland this bird finds the ground—or sea-level—temperature of the Arctic regions on the tops of the Grampians, and a few consequently remain behind to breed there. Numerous other instances might be given of northern species breeding in the south on high mountains, where the temperature is the same as that enjoyed in their usual habitat.

Though more strictly speaking a stranger of the spring, Pallas's Sand Grouse may still be not inaptly noticed in the present chapter. This bird wanders about a good deal in winter like the rest of the gipsy migrants, and appears to have no very regular migrations or true winter home. It is a bird of the steppes of Central Asia, and from time to time invades Europe in considerable "rushes." The last important visitation to the British Islands was in the spring of 1888; but it is more than doubtful whether this species will ever become naturalised in this country, in spite of the protection extended to it. It may readily be distinguished from all other British birds by its feathered legs and toes, the absence of a hind toe, and the pointed first primary and two centre tail feathers.

Y

CHAPTER V.

BEDTIME.

The habits of birds in winter are, perhaps, most interesting at the close of day. Most perching birds then seek shelter from the elements amongst evergreens, and the large shrubberies near fields and woods become suddenly alive with a regular rush of birds as soon as the sun draws near the western horizon. Every one who wishes to encourage birds should pay more than usual care to the cultivation of evergreens, as they form a perfect paradise for birds and for naturalists. It is in such places that birds are tempted to sing when they would not think of doing so in less sheltered localities. Evergreens furnish shelter from the severest storm, and warmth in the coldest weather — they are the grand refuge of nearly all the land birds that remain to winter in this country. To the dense, clustering shrubberies the birds repair at sundown, and during the fleeting hour of a winter twilight these places are the scene of many stirring incidents.

Of course, there are shrubberies and shrubberies. Those that are most favoured as the haunts of animated nature are the noble belts of evergreens round many of our English country seats—those ancient demesnes where the hollies and yews, laurels and myrtles, are a century old or more, and well sprinkled with a thick growth of deciduous underwood, amongst which are studded numerous oaks and elms, sycamores and ash trees, in which a rookery has generally been established for many generations. Memory recalls many such chosen spots, and doubtless the reader knows of others within his own experience —each and all the favoured retreat of bird life in winter. One of these noble shrubberies with which I am specially familiar stands on a gently sloping hillside in Derbyshire, almost within sight of the flaming furnaces of the Sheffield steelmakers. It is a wood and a shrubbery combined; tall trees with a thick undergrowth of evergreens and deciduous shrubs, such as hazel, elder, whitethorn, and sapling sycamores and elms — the whole forming a thick, dense cover and shelter, which all the year round is a great attraction for birds, but especially so in cold and severe weather. Amongst this luxuriant growth of vegetation the summer birds are in their glory. Shy Warblers rear their young where the bending hazels kiss the stream; Wood Wrens flit among the tree-tops; Spotted Flycatchers revel in the open glades. But now these summer migrants are far beyond the sea, the foliage in which they skulked is scattered to the winds, and the evergreens stand out in bold relief against the network of naked branches. Yet even now birds flock to this beautiful retreat from all the country-side; it is the chosen rendezvous, the grand headquarters of almost every bird in the immediate neighbourhood. Nightfall here in winter time is a busy, animated scene indeed. For many, many winters it has been my regular nightly pleasure to see all my feathered favourites safely to bed. I know almost every "bedroom" of every species in this highly favoured spot, where birds congregate in winter in such wonderful variety and in such enormous numbers. It is always an unqualified source of pleasure to me to see how widely birds appreciate the shelter of evergreens in winter; even an isolated yew tree or holly bush, far out in the open fields, will serve as a nightly resting-place for many a tired and weary songster; the tiniest of garden shrubberies will always entice birds in some numbers; then how much greater is the attraction of an extensive shrubbery such as this, where almost every species can find a haunt suited to its habits and requirements. No matter how deserted the fields and woods may be of bird life, or how unpropitious the weather for out-of-door observation of our feathered favourites, we are sure to find them in abundance among the evergreens, and their varied habits there are exceptionally engaging and full of interest.

Let us repair to this shrubbery to-night, and watch the actions of birds at the close of day. The sun, like a ball of dull, red fire, is settling down behind the distant snow-wreathed moors; fitful snowflakes whirl and eddy in the air as if foretelling another heavy fall; last night's hoar-frost is still encrusted on the grass; a cold wind rattles cheerlessly through the elm trees, sweeping across the open fields, penetrating all things with its pitiless, withering breath; everything presages an unusually cold night. But we shall find it warmer when we get among the sheltering shrubs. Out here scarcely a bird can be seen— the hedges are deserted, the fields are lonesome and dreary; now and then a tired and sleepy Finch flits overhead, twittering to itself as it goes, bending its course to the evergreens; we startle a few Blackbirds in the ditch by the meadow; they, too, hurry off with noisy cries to the old familiar roosting-place. But if the fields are deserted, we can hear the varied cries of birds in plenty from the shrubbery ahead; and as soon as we enter its gloomy portals we are among the birds in downright earnest at last.

Bitterly cold in the open though the evening is, the Robin's notes are sounding the day's requiem; and the glorious song of the Stormcock echoes high above the soughing of the wind. The speckled chorister is yonder, high up the bending elm, no bird that cleaves the air more wary than he. Rocked to and fro in the heightening gale, he pours out his evensong, then drops silently down into the hollies below, where later on we shall hear his rasping cry far into the twilight hour. In the hedgerows round the open fields not a Wren or a Hedge Sparrow gives voice, but here both of these little birds sing right heartily. It is astonishing what a little warmth and shelter will do in the matter of bird music. Birds like comfort just as much as we do; when they are comfortable, they are happy; and when they are happy, they express their feelings of joy in song and in calling to each other in a vast variety of tones.

The gloom of evening is now spreading over the woods and fields; already the evergreens begin to look black in the shadows, and birds on every side may be seen and heard settling down to rest amongst them. All the hard-billed birds, those that live on seeds and grain, are the first to retire to rest. They come here very early in the evening, and spend much time in twittering to each other, and flitting about the yews and hollies, searching out their sleeping quarters. All day long the beautiful Bramblings, Chaffinches, and Greenfinches have been busy on the neighbouring pastures, picking the seeds from the clover fields and stubbles, and the grain from the newly-sown land; now we see them flitting into the yew bushes, dropping down quickly from the elm and ash trees into the warm and welcome cover. The Bramblings often hold a friendly conversazione on the saplings before retiring to rest. These birds prefer the hollies, and are very socially inclined, many hundreds roosting within the area of a very few yards. The Chaffinches are more isolated in their tastes, and scatter themselves up and down the cover in yews, hollies, and myrtles indiscriminately, and often roost on the same twigs as Greenfinches. The Greenfinch loves the dense yew bushes better than any other evergreen, and in this respect shows the same partiality as the Bullfinch. Pairs of the latter bird steal silently up at nightfall, and glide softly into the yews. They have been busy all the day travelling up and down the weed-grown hedges, where dock and other seeds are plentiful; at dusk they draw near to the shrubberies, and we may hear their flute-like call-notes up and down amongst the trees. Companies of Titmice may here and there be seen flitting among the bare and leafless saplings. They, too, are wanderers, roaming the countryside all day in merry parties, seeking the shrubberies at dusk. Watch closely, and you will see them hop into the laurel and myrtle bushes one by one, quite silent now, and soon asleep. Then the watchful, wary Hawfinches, more like shadows than birds, glide into the yews with one or two clear-sounding good-night notes.

On every side the notes of birds are heard; some sonorous and loud, others low and softly-spoken; many shrill and piercing, a few harsh, discordant, and startling. Night is now coming on fast, the shadows deepen every moment, the wind sighs mournfully through the tree-tops high overhead. Bird after bird is continually arriving from the more open tracts of country. Now it is a pair of Ring Doves; these birds love the firs, and the rattle of their wings sounds startlingly clear as they dart quickly into their lofty apartments. Then a pair of noisy Jays come rollicking along—we heard them long ago in the distance; they are coming back to the old familiar holly tree in which they have slept every night all through the winter. A little later on the Magpies put in an appearance; they are rather late to-night, having extended their wanderings more than usual. How they chatter to each other in the pine trees! They, too, love a lofty roosting-place, and often sleep side by side. In the deepening gloom the Redwings come in a scattered flock. They prolong their stay upon the pastures until dusk, as is the way with all feeders on insects, worms, and other animal substances, although several pioneers of the vast assemblage arrived an hour ago, perhaps on fatigue duty, or acting the part of scouts. These birds are very noisy at the roosting-place, and perch for some considerable time on the bare saplings or in the tree-tops, chattering and calling to each other. As the darkness deepens bird after bird hops into the cover of the evergreens; but it will be observed that this species prefers to roost amongst the sycamore and elm saplings as long as any leaves remain upon them. Now in midwinter they select the holly by preference, though numbers roost in the yews, and a few in the laurels. On every side we can hear the noisy Blackbirds *pink-pink*ing to each other; and now and then the harsh cries of the Song Thrush, or the more discordant screams of the Stormcock sound with varying distinctness from the evergreens. Quarrels frequently take place for favourite corners, and for the best and most comfortable perches; all through the twilight the air resounds with a multitude of cries uttered in endless keys. The noisy chirp of the House Sparrow sounds incessantly from the ivy growing over the lofty sycamore and elm trees. This plant is preferred by the Sparrow before every other for roosting purposes, and the Wren is almost equally as fond of it. The delicate little Goldcrests also invariably roost among evergreens, coming into the shrubberies at sundown, after having spent the day among deciduous trees in the woods and hedges. From every part of the dark-looking evergreens we can hear the flutter of wings as sleepy birds settle down among the branches. The Robin still sings on into the deepening darkness, and his lovely strains are heard long after his form is lost in the gloom. Now and then, as if following the baton of some invisible leader, the songs and cries are hushed, and perfect quietness prevails for a few moments;

then the cries and flutterings are resumed. Sometimes a general uproar arises as the bold Sparrowhawk, hunting in the twilight for his supper, carries off some terror-stricken bird; or a clumsy Owl floats out of the clustering masses of ivy on his evening stroll in quest of food.

From time to time this shrubbery is the refuge of many distinguished and rare visitors, which call here in their wanderings, stay a night, and never return. Occasionally a small party of Crossbills steal quietly into the yew trees; and Twites and Siskins find shelter in the hollies and myrtles. Whilst standing carefully concealed under the evergreens, you may frequently catch a glimpse of these accidental visitors; whilst many of the regular sleepers here will pay you passing calls, or even settle themselves around you for the night. I have often had a Blackbird or a Redwing fly hurriedly up into the holly under which I was concealed, and, quite unconscious of my presence, settle itself to sleep. Last night a Blackbird came in this way, and after sitting on a branch for a quarter of an hour, calling at intervals to his companions near and far, I watched him hop on to a slender twig near the outside of the tree and settle down to rest. First he cleaned his bill by rubbing it sharply from side to side on the branch; then he shook his plumage, and listened intently for some moments, during which time I scarcely dared to breathe, and the beating of my heart was audible; and finally he tucked his yellow bill under the scapulary feathers, and, standing on one leg, quietly dropped off to sleep, looking twice his natural size, owing to the gloom and his puffed-out plumage.

As long as daylight lasts, birds of many species are constantly arriving, the evergreens forming the central point of attraction for all the birds in the neighbourhood. I am inclined to believe that many of these birds come here regularly to sleep, although they spend the daytime far away on the open fields, near the farmyards, and among the woods and coppices of deciduous trees. Then, as the darkness steals slowly over the earth, the bustle and excitement grow less and less until silence reigns supreme. Even the last noisy Wrens are silent, and the latest of the Robins has stayed its cheerful song. Thousands and thousands of birds are fast asleep around us; there is not one evergreen bush or tree that does not contain its feathered sleepers, warm and safe from every harm. Well may the naturalist prize the evergreens, for it is amongst them that wild life congregates in greatest variety and abundance.

There are some birds that seek shelter at nightfall in other places than shrubberies, to which we must devote half an hour at bedtime. Hay and corn-stacks and ivy growing over buildings shall be noticed first. Haystacks, especially those standing in the corners of fields, are the nightly resort of Titmice and Wrens. The Coal

Tit especially loves a haystack, and will return night after night to the accustomed roost. These birds generally roost just under the eaves formed by the overhanging thatch; but very often they make holes in the side to sleep in. The side opposite to the direction in which the wind is blowing is invariably selected, all the holes on the weather side being deserted. Very rarely indeed does more than one bird at a time sleep in the same hole. I have once or twice found a pair of Wrens sleeping in company, and Titmice less frequently still. The stacks in the farmyards are almost exclusively used by House Sparrows. Though Buntings, Chaffinches, and Greenfinches may be in these places in abundance during the day, it is rarely that they roost in the stacks, and never do so if evergreens are near at hand. Haystacks and those of oats are preferred to wheat, because in the latter the straw is hard and coarse. Vast numbers of Sparrows roost together in one stack, many repairing to them before sunset. All birds sleep very lightly, and the least noise usually startles them from their roosting-place, and you may hear bird after bird fluttering away into the darkness, dazed and bewildered. Ivy is another haunt of bird life; Sparrows, Blackbirds, Wrens, and Titmice are the most usual frequenters of this plant when growing over walls and buildings.

Many birds sleep in holes—sometimes those in which they rear their young, often in ones either burrowed out or selected ready made for the purpose. At dusk, the Woodpeckers seek a bedroom in the hollow trees, coming to certain holes night after night; and the Kingfisher retires at dusk to his cave under the banks of the stream. Herons and Rooks love to roost in fir trees. Pipits, Wagtails, and Larks sleep upon the ground, nestled among the dry herbage. Pheasants love the holly trees, but Partridges pass the night on the open fields.

Many birds there are that cannot avail themselves of the shelter of the evergreens. Some of them inhabit the wind-swept mountain-tops, where the breeze is ever sighing and moaning through the scanty herbage, and round the boulders and pebbles—the Ptarmigan, for example. But this bird crouches low amongst the crannies of the rocks, and, like the Red Grouse, lower down the hills, often buries itself in the snow, and sleeps secure from harm in the wreath when the weather is more than usually severe. The Eagles and the Falcons generally sleep at home among the cliffs where they make their nests; so, too, do the Crows and Ravens—all these are birds well able to stand the cold and the tempest; although one cannot help thinking that this constant exposure to the elements is one of the causes of their comparative rarity. Shore birds sleep much during the daytime, and are more or less alert and active during the night. Petrels sleep in their breeding holes; Gulls on the open banks, and on ocean rocks and islands; Divers and Auks most frequently on the sea; Cormorants in the caves and crannies of the cliffs. It will thus be seen that even in the manner of their roosting birds display considerable diversity of choice, and that their habits at eventide are replete with no ordinary interest.

CALENDAR FOR WINTER.

Species.	December.	January.	February.
Merlin.	Very rare	Very rare	Very rare
Kestrel.	Occasionally seen	Occasionally seen	More abundant
Missel-thrush.	In parties	In full song	Disperse to breeding grounds
Song Thrush.	Most have left	Returning	Returning
Redwing.	Remain in certain districts	Still in usual haunts	Still in usual haunts
Fieldfare.	Wandering about	Wandering about	Wandering about
Blackbird.	Most have left	Returning	Returning
Robin.	Draws near houses	Still in full song	Many pair
Stonechat.	In cultivated districts	In cultivated districts	In cultivated districts
Goldcrest.	In woods and hedges	Lives with Titmice	Lives with Titmice
Great Titmouse.	Wandering about	Wandering about	Utter love-notes
Blue Titmouse.	,,	,,	,,
Coal Titmouse.	,,	,,	,,
Marsh Titmouse.	,,	,,	,,
Long-tailed Titmouse	,,	,,	,,
Hedge Accentor.	Pairing	Pairing	Frequents breeding places
Wren.	Commonest in sheltered districts	Sings irregularly	,,
Creeper.	Seen with Titmice	Seen with Titmice	Seen with Titmice
Nuthatch.	,,	,,	,,

Species.	December.	January.	February.
Hooded Crow	On mud-flats	On mud-flats	On mud-flats
Rook	Visit nesting places	Visit nesting places	Begin to clean old nests
Waxwing	In parties now and then	In parties	In parties
Starling	Visits breeding holes	Visits breeding holes	Most have paired
Crossbill	Flocks roaming about	Flocks roaming about	Flocks roaming about
Bullfinch	Hedgerows & shrubberies	Hedgerows & shrubberies	Separate distinctly into pairs
House Sparrow	In stackyards	Shows signs of breeding	Few pairs begin nesting
Greenfinch	In large flocks	Still frequent fields	More often seen near shrubberies
Goldfinch	Wandering about	Wandering about	Wandering about
Siskin	,,	,,	,,
Brambling	On fields and in shrubberies	Attached to certain districts	Chatters on tree-tops
Chaffinch	Packs with Bramblings	Rapid abrasion of plumage	Pairing
Linnet	On weedy wastes	On weedy wastes	Still in flocks
Twite	,,	,,	,,
Lesser Redpole	,,	,,	,,
Snow Bunting	Often driven inland	Habitat according to weather	Many draw North
Yellow Bunting	In flocks on fields	In flocks on fields	Regain song

CALENDAR FOR WINTER.

Species.	December.	January.	February.
Pied Wagtail . . .	Occasionally seen	Occasionally seen	More frequently seen
Yellow Wagtail . .	,,	,,	,,
Meadow Pipit . .	Gregarious on lowlands	Gregarious on lowlands	Gregarious on lowlands
Skylark	On seeds and stubbles	On seeds and stubbles	Begin to disband
Shore Lark . . .	Seen on coast at intervals	Irregularly on coast	Draw North
Ring Dove . . .	Gregarious and silent	On fields, silent	On fields, silent
Ptarmigan	Descend to lower valleys	In full winter plumage	Begin disbanding
Heron	Most common on coast	On coast and estuaries	Congregate at breeding places
Moorhen	Leaves pools during frosts	Wanders little. Pairing	In pairs
Coot	In flocks on coast	In flocks on coast	In flocks on coast
Golden Plover . .	,,	,,	,,
Gray Plover . . .	Small numbers on coast	Small numbers on coast	Small numbers on coast
Lapwing	In flocks on lowlands	In flocks on lowlands	In flocks on lowlands
Knot	Flocks on coast	Flocks on coast	Flocks on coast
Dunlin	,,	,,	,,
Woodcock . . .	Inland woods, solitary	Inland woods, solitary	Rôding
Jack Snipe . . .	In swamps, solitary	In swamps, solitary	In swamps, solitary
Black-headed Gull .	On coast	On coast	On coast
Common Gull . .	Wandering about	Wandering about	Wandering about, and begin to moult

z

Species.	December.	January.	February.
Lesser Black-backed Gull	Wandering about	Wandering about	Wandering about, and begin to moult
Great Black-backed Gull	,,	,,	,,
Herring Gull . . .	,,	,,	,,
Kittiwake	,,	,,	,,
Auks	,,	,,	,,
Divers	Begin to assume nuptial plumage	Pairing	Generally at sea
Grebes	Most numerous near salt water	On and near coast	On and near coast
Geese	On and off coasts	On and off coasts	On and off coasts
Ducks	In nuptial dress	Pairing	Gatherings begin to disperse towards breeding grounds

INDEX.

Accentor, Alpine, points of distinction, 217.
Accentor, Alpine, The, 200.
Among the birds in Autumn, 169.
Among the birds in Winter, 272.
Among the wheat, 143.
Archeopteryx, The, 101.
Asiatic birds, 194.
Auk, Little, points of distinction, 228.
Auk, Little, The, 208.
Auks, Sleeping-places of, 333.
Autumn, Among the birds in, 169.
Autumn, Calendar for, 256.
Autumn, Features of the, 166.
Autumn, Song of birds in, 191.
Autumn, Strangers of the, 193.
Autumn, The beauties of the, 160.
Aviary, My ruined, 232.
Avocet, points of distinction, 50.
Avocet, The, 44, 46.
Azores and Bermudas, 195.

Basket Makers, 82.
Bass Rock, The, 33.
Beauty and utility, 52.
Bedtime, 322.
Bee Eater, Nest of, 76.
Bee Eater, points of distinction, 49.
Bee Eater, The, 43.
Belly, Pattern of colour on, 253.
Bermudas and Azores, 195.
Bird errantry, The philosophy of, 195.
Bird life, Disorganisation of, by frost, 273.

Bird music and warmth, 326.
Bird music in Winter, State of, 292.
Bird ornaments and tournaments, 52.
Birds, Affection of, for eggs and young, 137, 138.
Birds, Alluring artifices of, 129.
Birds, Arctic, 309.
Birds, Arts and tricks of, 125.
Birds, assumption of gregarious habits, 121.
Birds, Autumn migration of, 182.
Birds, Awakening of, 288.
Birds, Black cap or hood of, 251.
Birds, Colours of desert, 126.
Birds, Courtship of, 55, 56.
Birds, Crests of, 252.
Birds, Emotions of, 135.
Birds, Enjoyment of, in Autumn, 162.
Birds, Evening song of, 98.
Birds, Extension of range of, 196.
Birds, Feet of, 296, 297.
Birds, female, Tastes of, 58.
Birds, First broods of, 105.
Birds, Gatherings of, on coast, 236.
Birds, Habits of, at dawn in Summer, 94.
Birds, hard-billed, Habits of, at evening, 326.
Birds, High mortality among, 165.
Birds, Language of, 138.
Birds, Light sleep of, 332.
Birds, loss of song in Summer, 118.
Birds, Love song of, 56.
Birds, Mating of, 140, 141.
Birds, Mental attributes of, 23.

Birds, Mental qualities of, 139, 142.
Birds, Migration of, 14.
Birds, Migration of, in Winter, 277.
Birds, Moulting of, 120.
Birds, Movements of, in Autumn, 167.
Birds, Notes of, at dusk, 327, 328, 329.
Birds, Number of species of British, 193.
Birds, Patience of, 139.
Birds, Perseverance of, 139.
Birds, Power of conversing of, 136.
Birds, Raptorial, Aërial evolutions of, 66.
Birds, Shamming death by, 128.
Birds, sleeping in company, 332.
Birds, Sociability of, 140.
Birds, soft-billed, Services rendered by, 150, 151.
Birds, Song of, in Spring, 24, 36.
Birds, Songs of, 140.
Birds, Storm-driven, 274.
Birds, Sympathy of, 137.
Birds, The banishment of the, 231.
Birds, The ways of, 134.
Birds, Usefulness of, 143.
Birds, Various notes of, 136, 138.
Birds, Wings of, 253.
Birds, Young, 37.
Birds, young, Flocks of, 172.
Bittern, American, points of distinction, 224.
Bittern, American, The, 207.
Bittern, Little, points of distinction, 50.
Bittern, Little, The, 44.
Bittern, The, 182.
Blackbird, Filaments on head of, 253.
Blackbird, Incident of, at night, 330.
Blackbird, Tracks of, in snow, 303, 304.
Blackbird, Winter movements of, 277.
Blackbirds, Young of, 123.
Blackcap, Arrival of, 16.
Blackcap, The, 46.
Blackcap, Winter home of, 242.
Blackcock, Pugnacity of, 58.
Bluethroat, points of distinction, 215, 216.

Brambling, Arrival of, 177.
Brambling, Habits of, 284, 285, 327.
Bramblings, Twitterings of, 291.
Broads, Autumn evenings on, 181.
Broads, The, 181.
Bulbuls, Pattern of colour in, 253.
Bullfinch, Eastern form of the, 202.
Bullfinch, Eastern, points of distinction, 219.
Bullfinches in Winter, 284.
Bullfinch, Nesting of, 28.
Bunting, Black-headed, points of distinction, 221.
Bunting, Black-headed, The, 203.
Bunting, Lapland, points of distinction, 220.
Bunting, Lapland, The, 203.
Bunting, Little, points of distinction, 220.
Bunting, Little, The, 203.
Bunting, Ortolan, points of distinction, 221.
Bunting, Ortolan, The, 203.
Bunting, Rustic, points of distinction, 221.
Bunting, Rustic, The, 203.
Bunting, Snow, Departure of, 14.
Bunting, Snow, Distribution of, 317, 318.
Bunting, Snow, Food of, 319.
Bunting, Snow, Habits of, 318.
Bunting, Snow, Nest and eggs of, 319, 320.
Bunting, Snow, Plumage of, 319.
Buntings, Services rendered by, 149.
Bunting, Yellow, Fecundity of, 120.
Bustard, Macqueen's, 206.
Bustard, Macqueen's, points of distinction, 224.
Buzzard, Honey, points of distinction, 47.
Buzzard, Honey, Winter home of, 241.
Buzzard, Rough-legged, Migration of, 184.
Buzzard, Rough-legged, Winter home of, 246.

INDEX. 341

Calendar for Autumn, 256.
Calendar for Winter, 335.
Canary, points of distinction, 219.
Canary, Wild, The, 202.
Capercailzie, Tracks of, 307.
Cards, Some visiting, 296.
Cedar Bird, 310.
Chaffinch, Assumption of breeding plumage, 293.
Chaffinches, Arrivals of, 184.
Chaffinches in Autumn, 177.
Chaffinches, Roosting habits of, 327.
Chaffinch, Nest of, 18, 78, 79.
Chaffinch, Pairing of, 293.
Chat, Black-throated, points of distinction, 216.
Chat, Black-throated, The, 199.
Chat, Isabelline, points of distinction, 216.
Chat, Isabelline, The, 199.
Chats, Characteristics of plumage in, 251.
Chiffchaff, Courtship of, 5.
Chiffchaff, Departure of, 177.
Chiffchaff, Migrations of, 4.
Chiffchaff, Nest-building of, 6.
Chiffchaff, Song of, 12.
Chiffchaff, Winter home of, 242.
Chin and throat patterns of colour, 252.
Coast Birds, changes among, in Autumn, 179.
Cold weather, Scarcity of some birds in, 278.
Colour, Patterns of, 249.
Coot, Footprints of, 300.
Coot, Nest of, 84.
Coots, Migratory, 282.
Cormorant, Footprints of, 299.
Cormorants, Nests of, 82.
Cormorants, Nuptial ornaments of, 53.
Cormorants, Sleeping-places of, 333.
Corn-crake, Arrival of, 16.
Corn-crake, Winter range of, 243.
Corn-stacks as roosting-places, 332.

Courser, Cream-coloured, points of distinction, 224.
Courser, Cream-coloured, The, 206.
Crake, Little, points of distinction, 211.
Crake, Little, The, 210.
Crane, Demoiselle, points of distinction, 50.
Crane, Demoiselle, The, 44.
Crane, points of distinction, 50.
Crane, The, 44.
Creepers and Nuthatches, 234.
Creeper, Wall, points of distinction, 211.
Creeper, Wall, The, 210.
Crossbill, American, points of distinction, 219.
Crossbills, Habits of, at nightfall, 330.
Crossbill, The, 279.
Crossbill, White-winged, American form of, 201.
Crossbill, White-winged, points of distinction, 219.
Crossbill, White-winged, The, 201.
Crow, Carrion, Nesting of, 21.
Crow, Hooded, Arrival of, 179.
Crows, Gatherings of, 67.
Crows, Habits of, 65.
Crows, Nests of, 82.
Crows, Roosting-place of, 333.
Cuckoo, Arrival of, 15.
Cuckoo, change of note, 105.
Cuckoo, date of laying, 30.
Cuckoo, Departure of, 172.
Cuckoo, Great Spotted, points of distinction, 49.
Cuckoo, Great Spotted, The, 43.
Cuckoo, Habits of, 15.
Cuckoo, Mating of, 141.
Cuckoo, Resemblance of, to Hawks, 126.
Cuckoo, Winter home of, 245.
Cuckoo, Yellow-billed, points of distinction, 223, 224.
Cuckoo, Yellow-billed, The, 207.
Cuckoo, Young of, 117.

Curlew, Esquimaux, points of distinction, 225.
Curlew, Esquimaux, The, 207.
Curlews, Young, on coast, 123.

Deceit, Practice of, 124.
Derbyshire shrubbery, A, 323.
Dipper, Black-bellied, points of distinction, 215.
Dipper, Habits of, 62.
Dipper, Habits of, in Winter, 279.
Dipper, Nest of, 19, 79.
Dipper, Scandinavian form of, 198.
Diver, Great Northern, Habits of, in Winter, 294.
Divers, Sleeping-places of, 333.
Divers, Wedding garments of, 54.
Diver, White-billed, points of distinction, 228.
Diver, White-billed, The, 208.
Dotterel, Departure of, 183.
Dotterel, Winter home of, 244.
Dotterel, Wonderful migrations of, 183.
Dove, Ring, Fecundity of, 120.
Dove, Ring, Footprints in snow of, 303.
Dove, Ring, Habits of, in Autumn, 238.
Dove, Stock, Habits of, in Autumn, 238.
Dove, Stock, on moors in Autumn, 235.
Dove, Turtle, Arrival of, 17.
Dove, Turtle, Departure of, 177.
Duck, Buffel-headed, points of distinction, 229.
Duck, Buffel-headed, The, 210.
Ducks, Date of pairing of, 295.
Ducks, Gatherings of, 237.
Ducks, Moulting of, 120.
Ducks, Plumage of, 255.
Duck, Tufted, The, 237
Ducks, Wedding garments of, 54.
Duck, Wild, Breeding of, 35.
Dunlin, date of laying, 30.
Dunlin, Return of, to coast, 121.
Dunlins, Arrival of, 179.

Eagle, Golden, Nest of, 63.
Eagle, Spotted, points of distinction, 214.
Eagle, Spotted, The, 197, 198.
Eagles, Roosting-place of, 333.
Eagle, White-tailed, The, 68.
Egret, Great White, points of distinction, 49.
Egret, Great White, The, 44.
Egret, Little, 44.
Egret, Little, points of distinction, 49.
Eider Duck, Nest of, 85.
Eider, Steller's, 210.
Eider, Steller's, points of distinction, 230.
Elephant, Slow increase of, 164.
Engineers, Our feathered, 70.
Evening in Summer, 98.
Evergreens, Among the, 331.
Evolution, Thoughts on, 100.

Falcon, Brown Jer, points of distinction, 214.
Falcon, Iceland Jer, points of distinction, 214.
Falcon, Red-footed, points of distinction, 213.
Falcon, Red-footed, The, 197.
Falcons, Jer, The, 197,
Falcons, Roosting-place of, 333.
Falcon, White Jer, points of distinction, 214.
Farne Islands, The, 32.
Feathered frauds, 124.
Feathers, Modification of, 57.
Felt Makers, 78.
Fieldfare, Arrival of, 181.
Fieldfares, Departure of, 14.
Fields and woods, Autumn aspect of, 161.
Finches, Abrasion of plumage of, 54.
Finches, Autumn flights of, 233.
Finches, Habits of, in Winter, 275.
Finches, Services rendered by, 149.

INDEX.

343

Finch, Scarlet Rose, points of distinction, 219.
Finch, Scarlet Rose, The, 202.
Firecrest, points of distinction, 217.
Firecrest, The, 200.
Flycatcher, Gray, Departure of, 176.
Flycatcher, Pied, Nest of, 78.
Flycatcher, Pied, Winter home of, 243.
Flycatcher, Red-breasted, points of distinction, 216, 217.
Flycatcher, Red-breasted, The, 199.
Flycatcher, Spotted, Abundance of, in Northern Africa, 245.
Flycatcher, Spotted, Arrival of, 17.
Flycatcher, Spotted, Habits of, 23.
Flycatcher, Spotted, Nest of, 103.
Flycatcher, Spotted, Parties of, 122.
Flycatcher, Spotted, Winter home of, 244.
Footprints in the farmyard, 306.
Footprints on the mud-flats, 297.
Frauds, Feathered, 124.
Frost, A long-continued, 268.
Fruit trees, Blossom of, destroyed by Sparrows, 148.
Fulmar, date of laying, 32.
Fulmar Petrel, Breeding habits of, 113.

Gannet, Breeding habits of, 34.
Gannet, date of laying, 34.
Gannet, Note of, 34.
Gannets, 33.
Garganey, points of distinction, 51.
Garganey, The, 45.
Geese, Flight of, in Winter, 276.
Geese, Moulting of, 120.
Geese, Nests of, 85.
Geese, Snow, 209.
Geese, Snow, points of distinction, 228.
Geese, Tracks left by, 298.
Geographical distribution, Peculiarities of, 40.
Gipsy migrants, 279.
Glacial epoch, Birds during last, 195.
Goatsucker, Arrival of, 17.

Goatsuckers, Plumage of, 255.
Goatsucker, Winter home of, 244.
Goldcrests, Autumn flights of, 176.
Goldcrests, Disbanding of, 276.
Goldcrests, Habits of, 176.
Golden-Eye, Nest of, 85.
Goldfinches, 284.
Goldfinches, Habits of, in Autumn, 188.
Goose, Red-breasted, points of distinction, 229.
Goose, Red-breasted, The, 209.
Goshawk, American, points of distinction, 214.
Goshawk, American, The, 198.
Goshawk, points of distinction, 214.
Goshawk, The, 198.
Grebe, Black-necked, points of distinction, 212.
Grebe, Black-necked, The, 210.
Grebe, Great Crested, The, 181.
Grebe, Little, Tracks of the, 304.
Grebes, Nests of, 84.
Greenfinches, Roosting-places of, 327.
Greenfinch, Fecundity of, 120.
Greenfinch, Nesting of, 28.
Grosbeak, Pine, Bill of, 316.
Grosbeak, Pine, Distribution of, 315, 316.
Grosbeak, Pine, Food of, 316.
Grosbeak, Pine, Habits of, 316.
Grosbeak, Pine, Nest and eggs of, 317.
Grosbeak, Pine, points of distinction, 218.
Grosbeak, Pine, The, 201.
Grouse, Black, Tracks of, 307.
Grouse, Pallas's Sand, 321.
Grouse, Red, burrowing in snow, 282.
Grouse, Red, Comb of, 53.
Grouse, Red, Footprints of, 307.
Grouse, Red, Notes of, 29.
Grouse, Red, Protective colour of, 127.
Grouse, Red, Roosting-place of, 333.
Guillemot, date of laying, 32.
Gull, Black-headed, Habits of, in Autumn, 190.

Gull, Glaucus, points of distinction, 227.
Gull, Glaucus, The, 207.
Gull, Iceland, points of distinction, 227.
Gull, Iceland, The, 207.
Gull, Ivory, points of distinction, 227.
Gull, Ivory, The, 207.
Gull, Little, points of distinction, 227.
Gull, Little, The, 181, 207.
Gull, Ross's, 207.
Gull, Ross's, points of distinction, 227.
Gull, Sabine's, 207.
Gull, Sabine's, points of distinction, 227.
Gulls and Terns, Gregarious instincts of, 238.
Gulls, date of laying, 33.
Gulls, Footprints of, 301.
Gulls, Moulting of, 120.
Gulls, Sleeping-places of, 333.
Gulls, Spring plumage of, 54.

Hang-Nests, Nests of, 81.
Happiness in Nature, 163.
Harrier, Hen, points of distinction, 47.
Harrier, Hen, Winter home of, 241.
Harrier, Marsh, Winter home of, 241.
Harrier, Montagu's, 39.
Harrier, Montagu's, points of distinction, 47.
Harrier, Montagu's, Winter quarters of, 241.
Hawfinches, Gatherings of, in Winter, 285, 286.
Hawfinches, Roosting-places of, 327.
Hawk, Sparrow, Nesting of, 27.
Hawks, Usefulness of, 147, 148.
Hay meadows, Summer in the, 109.
Haystacks as roosting-places, 331.
Hedge Sparrow, Nest of, 18.
Heligoland, 194, 195.
Hen Harrier, The, 39.
Heron, Buff-backed, points of distinction, 49, 50.
Heron, Buff-backed, The, 44.
Heron, Eggs of, 25.

Heron, Footprints of, 299.
Heron, Nesting of, 24.
Heron, Night, points of distinction, 50.
Heron, Night, The, 44.
Heron, Purple, points of distinction, 224.
Heron, Purple, The, 205.
Herons, Habits of young, in Autumn, 188.
Herons, Nuptial ornaments of, 54.
Heron, Squacco, points of distinction, 49.
Heron, Squacco, The, 44.
Herons, Roosting-place of, 333.
Hobby, 22.
Hobby, points of distinction, 47.
Hobby, The, 39.
Hobby, The Winter quarters of the, 241.
Holes as roosting-places, 332, 333.
Honey Buzzard, The, 39.
Hooded Crow, Nest of, 68.
Hoopoe, points of distinction, 49.
Hoopoe, The, 43, 46.

Ibis, Glossy, points of distinction, 224.
Ibis, Glossy, The, 205.
Ivy as a haunt of birds, 332.

Jay, Nest of, 21.
Jays and Magpies in Winter, 276.
Jay, Siberian, Plumage of, 255.
Jays, Parties of, in Summer, 121.
Jays, Roosting-place of, 328.

Kestrel, Habits of, 111.
Kestrel, Lesser, points of distinction 214.
Kestrel, Lesser, The, 197.
Kestrel, Movements of, in Winter, 277.
Kestrel, Nesting of, 27.
Kestrel, Winter home of, 241.
Kingfisher, Belted, points of distinction, 223.
Kingfisher, Belted, The, 207.
Kingfisher, Footprints of, 305.
Kingfisher, Habits of, in Winter, 279, 280.

INDEX.

Kingfisher, Migrations of, 190.
Kingfisher, Nest of, 76.
Kingfisher, young, Distinctive characters of, 190.
Kite, Black, points of distinction, 47.
Kite, Black, The, 40.
Kite, Swallow-tailed, points of distinction, 215.
Kite, Swallow-tailed, The, 198.
Kittiwake, date of laying, 33.
Kittiwake, Nest of, 72.
Knots, arrival of old birds, 179.
Knots, Arrival of, on coast, 123.
Knot, Winter home of, 246.

Landrail, Cessation of note of, 112.
Landrail, Departure of, 175.
Landrail, Habits of, 112.
Landrail, Nest of, 110.
Landrail, Shamming death by, 128.
Lapwing, Alluring artifices of, 130.
Lapwings, Flocks of, in Winter, 274.
Lapwings, Habits of, in Autumn, 190.
Lark, Calandra, points of distinction, 223.
Lark, Calandra, The, 204.
Lark, Crested, points of distinction, 222.
Lark, Crested, The, 204.
Lark, Desert, 126.
Larks and Buntings in Winter, 277.
Larks, Eggs of the, 315.
Lark, Shore, Distribution of, 313.
Lark, Shore, Economy of, 314.
Lark, Shore, Nest and eggs of, 315.
Lark, Shore, points of distinction, 223.
Lark, Shore, Pre-glacial ancestor of, 313.
Lark, Shore, The, 205.
Lark, Shore, Young of, 315.
Lark, Short-toed, points of distinction, 222.
Lark, Short-toed, The, 204.
Larks, Roosting-place of, 333.
Lark, White-winged, points of distinction, 222, 223.

Lark, White-winged, The, 204.
Lark, Wood, Change of habits in, 293.
Lark, Wood, Song of, 293.
Lesser Redpoles, 234.
Lighthouses and migration, 185.
Lincolnshire, Snow Bunting in, 318.
Linnet, Breeding of, 30.
Linnet, Disbanding of, 13.
Local migrations in Winter, 278.

Magpie, Nest of, 20, 82, 86.
Mapies, Parties of, in Summer, 121.
Magpies, Roosting-place of, 328.
Martin, House, Arrival of, 15.
Martin, Purple, points of distinction, 221.
Martin, Purple, The, 207.
Martin, Sand, Arrival of, 14.
Martin, Sand, Nest of, 74, 75.
Martins, Departure of, 172.
Martins, Nest of, 23, 73.
Masons, 72.
Mental capabilities of lower animals, 2.
Merganser, Hooded, points of distinction, 230.
Merganser, Hooded, The, 210.
Merganser, Nest of, 85.
Merlin, Autumn movements of, 172.
Merlin, Nesting of, 30.
Midday in Summer, 95, 96.
Migrants, Habits of, abroad, 247.
Migrants regaining song, 247.
Migrants, Routes followed by, 246.
Migration, Autumn, 194.
Migration, Facts about, 184.
Migration in Spring, 39.
Migration, The fraternity of, 186.
Migration, The perils of, 184.
Milder weather, 269.
Miners, 75.
Missel-thrush, Autumn song of, 170.
Missel-thrush, Habits of, in Summer, 113.
Missel-thrush, Song of, in Winter, 290.
Mixed congregations, 231.
Moorhen, Footprints of, 304.

Moorhen, Nest of, 83.
Moorhen, Nocturnal habits of, in Winter, 282.
Moorlands in Autumn, 235.
Moorlands, The, in Winter, 282.

Naturalist, Pleasures of the, 8.
Nature, Lavishness of production in, 165.
Nest, House Sparrow's, double type of, 27.
Nestlings, Tricks and deceit practised by, 130.
Nests, 18.
Nests, Concealment of, 129.
Netting birds on the Wash, 180.
Nidification and temperature, 321.
Nightingale, Arrival of, 16.
Nightingale, Departure of, 177.
Nightingale, Song of, 17.
Night in the shrubbery, 325.
Nightjar, date of laying, 111.
Nightjar, Departure of, 178.
Nightjar, Habits of, 111.
Nightjar, Isabelline, points of distinction, 48.
Nightjar, Isabelline, The, 42.
Nightjar, Red-necked, points of distinction, 48, 49.
Nightjar, Red-necked, The, 42.
Nile valley, Migration down the, 246.
Nutcracker, points of distinction, 217.
Nutcracker, The, 201.
Nuthatches and Creepers, 234.
Nuthatch, Nest of, 72.

Oases, Beauty of the, 243.
October, Bird life in, 188.
Old breeding-places, Birds' visits to, 289.
Organic life, Mortality among, 164.
Organs emphasized on plumage, 254.
Oriole, Golden, Nest of, 81.
Oriole, Golden, points of distinction, 48.
Oriole, Golden, The, 41.
Ornithology, The pleasures of, 134.

Osprey, Winter home of, 241.
Ousel, Black-throated, points of distinction, 215.
Ousel, Black-throated, The, 198.
Ousel, Ring, Departure of, 171.
Ousel, Ring, Flocking of, 122.
Ousel, Ring, Nesting of, 29.
Ousel, Ring, Winter home of, 241.
Ousel, Ring, Young of, 117.
Ousels, Streaked throat of, 250.
Owl, Hawk, points of distinction, 215.
Owl, Hawk, The, 198.
Owl, Scops, points of distinction, 47.
Owl, Scops, The, 40.
Owl, Short-eared, Migrations of, 175.
Owl, Snowy, points of distinction, 215.
Owl, Snowy, The, 198.
Owls, Plumage of, 255.
Owls, Services rendered by, 152, 153.
Owl, Tengmalm's, 198.
Owl, Tengmalm's, points of distinction, 215.
Oystercatcher, date of laying, 35.
Oystercatcher, Eggs of, 35.
Oystercatcher, Footprints of, 300.

Palæarctic birds, Eastern range of, 195.
Paragraphs on plumage, 249.
Partridge, Habits of, 174.
Partridge, Pairing of, 293.
Partridge, Roosting-place of, 333.
Partridges, Broods of, 106, 107.
Partridges, Rock, Colour of throat in, 252.
Pastor, Rose-coloured, points of distinction, 218.
Pastor, Rose-coloured, The, 201.
Peewits, 29.
Peewit, Wing of, 59.
Peregrine, Nest of, 65.
Peregrine, Swoop of, 66.
Petrel, Fork-tailed, Date of laying and nesting habits of, 115.
Petrels, Sleeping-places of, 333.
Petrels, Southern wanderings of, 180.

INDEX. 347

Petrel, Stormy, Nesting habits of, 115.
Petrel, Wilson's, points of distinction, 51.
Petrel, Wilson's, 45.
Phalarope, Gray, points of distinction, 224, 225.
Phalarope, Gray, The, 206.
Pheasant hen, Protective colour of, 127.
Pheasants, Feeding of, on acorns, 176.
Pheasants, Roosting-place of, 333.
Pheasant, The comb of, 53.
Pheasant, Tracks of the, 303.
Pied Wagtails, Migration of, 13.
Pigeons, Nests of, 83.
Pigeons, Plumage of, 255.
Pinnacles, The, 32.
Pipit, Alpine, points of distinction, 222.
Pipit, Alpine, The, 204.
Pipit, Meadow, Habits of, in Winter, 285.
Pipit, Meadow, Nesting of, 30.
Pipit, Meadow, Return of, to lowlands, 171.
Pipit, Red-throated, points of distinction, 221, 222.
Pipit, Red-throated, The, 203.
Pipit, Richard's, points of distinction, 222.
Pipit, Richard's, 204.
Pipits, Roosting-place of, 333.
Pipit, Tawny, points of distinction, 222.
Pipit, Tawny, The, 204.
Pipit, Tree, Arrival of, 24.
Pipit, Tree, Broods of, 112.
Pipit, Tree, Departure of, 177.
Pipit, Tree, Nest and eggs of, 24.
Pipit, Tree, Winter home of, 242.
Plasterers, 71.
Ploughed land, Birds on, 189.
Plover, American Golden, points of distinction, 225.
Plover, Asiatic Golden, points of distinction, 225.
Plover, Asiatic Golden, The, 206.
Plover, Golden, Colours of, 253.
Plover, Golden, Nesting of, 29.
Plover, Kentish, Winter range of, 243.

Plover, Killdeer, points of distinction, 225.
Plover, Killdeer, The, 207.
Plover, Ringed, Eggs of, 35.
Plover, Ringed, Nesting habits of, 131, 132.
Plover, Ringed, Winter range of, 243.
Plover, Spur-winged, The, 59.
Plumage, Paragraphs on, 249.
Plumage, Peculiarities of, 249.
Plumage, Texture of, 254.
Pochard, Red-crested, points of distinction, 229.
Pochard, Red-crested, The, 210.
Pochard, The, 237.
Pochard, The, on the Broads, 181.
Pratincole, points of distinction, 211, 212.
Pratincole, The, 210.
Ptarmigan, date of hatching, 117.
Ptarmigan in Winter, 282.
Ptarmigan, Protective colour of, 127.
Ptarmigan, Roosting-place of, 333.
Ptarmigan, Tracks of, 307.
Puffin, Nest of, 76.
Puffin, Return of, to St. Kilda, 14.

Quail, Nest of, 107.
Quail, Winter range of, 243.

Rabbit, Death of, in snow, 307.
Raft Makers, 83.
Raven, Nest of, 64, 65.
Ravens, Roosting-place of, 333.
Razorbill, date of laying, 32.
Red Grouse, Footprints of, 307.
Red Grouse, Packing of, 123.
Redpole, Greenland, points of distinction, 220.
Redpole, Greenland, The, 202.
Redpole, Lesser, Eggs of, 22.
Redpole, Mealy, points of distinction, 220.
Redpole, Mealy, The, 202.
Redpole, Nesting of, 104.
Redpoles, Disbanding of, 13.
Redshank, Dusky, points of distinction, 212.

INDEX

Redshank, Dusky, The, 210.
Redstart, Departure of, 177.
Redstart, Nest of, 78, 103.
Redstarts, Characteristics of plumage in, 251.
Redstart, Winter home of, 243.
Redwings, Arrival of, 177.
Redwings, Departure of, 14.
Redwings, Roosting habits of, 328.
Redwings, Sufferings of, in hard weather, 275.
Ring Doves, Cooing of, 13.
Ring Doves, Roosting of, 328.
Robin, Autumn song of, 169.
Robin, Beauty of Autumn song of, 187.
Robin, Nest of, 18.
Robin, Pugnacity of, 57.
Robin, regaining of song, 121.
Robin, Winter song of, 290.
Rock birds, Habits of, in Winter, 294.
Roller, points of distinction, 211.
Roller, The, 210.
Rook, Interesting habits of, in Autumn, 178.
Rook, Nesting of, 20.
Rooks and Starlings in Winter, 277.
Rooks, cleaning out nests, 289.
Rooks feeding on acorns, 176.
Rooks, Footprints of, 304, 306.
Rooks, Roosting-place of, 333.
Rooks, services to man, 152.
Rooks, Visits of, to nest trees, 289.
Ruff, Collar shield of, 54.
Ruff, Combats of, 59.
Rump, Pattern of colour on, 253.

Sanderling, The, 236.
Sanderling, Tracks made by, 300.
Sandpiper, Bartram's, 207.
Sandpiper, Bartram's, points of distinction, 225.
Sandpiper, Bonaparte's, 207.
Sandpiper, Bonaparte's, points of distinction, 226.
Sandpiper, Broad-billed, points of distinction, 212.

Sandpiper, Broad-billed, The, 210.
Sandpiper, Buff-breasted, points of distinction, 227.
Sandpiper, Buff-breasted, The, 207.
Sandpiper, Common, Alluring artifices of, 129.
Sandpiper, Common, Departure of, 188.
Sandpiper, Common, Habits of, 105, 106.
Sandpiper, Common, Winter range of, 243.
Sandpiper, Curlew, The, 236.
Sandpiper, Green, points of distinction, 212.
Sandpiper, Green, The, 210.
Sandpiper, Pectoral, points of distinction, 226.
Sandpiper, Pectoral, The, 207.
Sandpiper, Purple, The, 236.
Sandpiper, Solitary, points of distinction, 225.
Sandpiper, Solitary, The, 207.
Sandpiper, Spotted, points of distinction, 225.
Sandpiper, Spotted, The, 207.
Sandpiper, Wood, points of distinction, 212.
Sandpiper, Wood, The, 210.
Sandpiper, Yellow-legged, points of distinction, 226.
Sandpiper, Yellow-legged, The, 207.
Scaffold Builders, 83.
Scoter, Surf, points of distinction, 230.
Scoter, Surf, The, 210.
Sedge Bird, Descent of, 100.
Serin, points of distinction, 219.
Serin, The, 202.
Sexes, Separation of, during migration, 248.
Shearwater, Dusky, points of distinction, 51.
Shearwater, Dusky, The, 45.
Shearwater, Manx, date of laying, 114.
Shearwater, Sooty, points of distinction, 228.
Shearwater, Sooty, The, 208.

INDEX.

Sheldrake, Ruddy, points of distinction, 229.
Sheldrake, Ruddy, The, 209.
Shore, Bird life on, in Winter, 281.
Shore birds, Arrival of, 180.
Shore birds, Sleeping-places of, 333.
Shrike, Great Gray, points of distinction, 218.
Shrike, Great Gray, The, 201.
Shrike, Lesser Gray, points of distinction, 211.
Shrike, Lesser Gray, The, 210.
Shrike, Pallas's Gray, points of distinction, 218.
Shrike, Pallas's Gray, 201.
Shrike, Red-backed, Curious migrations of, 245.
Shrike, Red-backed, Nesting of, 103.
Shrike, Red-backed, Winter home of, 245.
Shrikes, Moulting of, 248.
Shrike, Woodchat, points of distinction, 48.
Shrike, Woodchat, The, 41.
Shrubberies, 323.
Shrubberies in Spring, 28.
Shrubbery, Nightfall in the, 324.
Siskin, Nest of, 22.
Siskins, 234.
Siskins, Habits of, in Autumn, 187.
Siskins, Roosting-place of, 330.
Skua, Buffon's, points of distinction, 228.
Skua, Buffon's, The, 208.
Skua, Pomarine, points of distinction, 228.
Skua, Pomarine, The, 208.
Skylark, Autumn flocking of, 171.
Skylark, Autumn song of, 170, 171.
Skylark, Flocking of, 112.
Skylark, Song of, 24.
Skylarks, Tracks left by, 304.
Skylark, Winter song of, 291.
Small birds, Migrations of, 184.
Snipe, Drumming of, 28.
Snipe, Eggs of, 29.
Snipe, Footprints of the, 305.

Snipe, Jack, Arrival of, 183.
Snipe, Jack, Habits of, in Winter, 280.
Snipe, Jack, Migrations of, 28.
Snipe, Red-breasted, points of distinction, 226.
Snipe, Red-breasted, The, 207.
Snipe, Sounds made by, 57.
Snipes, Protective colour of, 127.
Snipes, Solitary habits of, 281.
Snowstorm, The grandeur of the, 264.
Snow, The first fall of, 264.
Snow, Writing on the, 302.
Song of birds in Autumn, 191.
Song of birds in Spring, 36.
Song of birds in Summer, 119.
Song of birds in Winter, 286, 290.
Some birds of the Winter, 309.
Some visiting cards, 296.
Sparrow Hawk, hunting in the gloom, 330.
Sparrow Hawk, Marks left by, 307.
Sparrow, Hedge, Autumn song of, 170.
Sparrow, Hedge, Late broods of, 119.
Sparrow, Hedge, Moulting of, 283.
Sparrow, House, Destructive habits of, 145.
Sparrow, House, Early nesting of, 293.
Sparrow, House, Food of, 146, 147.
Sparrow, House, Loss caused by, 146.
Sparrow, House, Nesting of, 26.
Sparrow, House, Nest of, 80, 81.
Sparrow, House, services to man, 144, 146.
Sparrows, Favourite roosting-places of, 329.
Sparrows, Hedge, migratory, 283.
Spoonbill, points of distinction, 50.
Spoonbill, The, 44, 46.
Spring, Among the birds in, 10.
Spring, Calendar for, 87.
Spring, First signs of, 292.
Spring, Migration of birds in, 39.
Spring, Strangers of the, 38.
Spring, The glories of the, 1.
Spring-time on the mountains, 61.
Starling, Autumn song of, 170.

Starling, Footprints in snow of, 304.
Starling, Habits of, in Autumn, 186, 187.
Starling, Habits of, in Summer, 112.
Starlings, services to man, 152.
Starlings, Tracks of, 306.
Starling, young, Song of, 187.
Stilt, Black-winged, points of distinction, 50, 51.
Stilt, Black-winged, The, 44.
Stint, American, points of distinction, 226.
Stint, American, The, 207
Stint, Temminck's, points of distinction, 212.
Stint, Temminck's, 210.
St. Kilda, 113.
Stonechat, date of laying, 30.
Stonechat, Habits of, in Winter, 285.
Stone Curlew, Departure of, 183.
Stork, Black, points of distinction, 211.
Stork, Black, The, 210.
Stork, White, points of distinction, 211.
Stork, White, The, 210.
Stormcock, Evensong of, in Winter, 325, 326.
Strangers of the Autumn, 193.
Stream, Banks of the, in Winter, 279.
Struggle for existence, The, 163.
Stubbles in Autumn, 173.
Stubbles in Winter-time, The, 276.
Summer, Among the birds in, 102.
Summer, Calendar for, 154.
Summer, Flowers of, 97.
Summer, Night-time in, 99.
Summer, Reverie in the woods in, 99.
Summer, The wonders of the, 93.
Swallow, Arrival of, 14.
Swallow, Departure of, 172.
Swallows, Congregation of, 172, 173.
Swallows, Moulting time of, 121.
Swallows, Nests of, 22, 73.
Swallow, Song of, 111.
Swallow, Winter home of, 245.
Swan, Bewick's, 209.
Swan, Bewick's, points of distinction, 228.

Swans, Footprints of, 298, 299.
Swift, Alpine, points of distinction, 48.
Swift, Alpine, The, 42.
Swift, Arrival of, 17.
Swift, date of laying, 106.
Swift, departure of old birds, 178.
Swift, Needle-tailed, points of distinction, 223.
Swift, Needle-tailed, The, 205.
Swifts, young, Departure of, 123.
Swift, Winter home of, 244.

Tailor Bird, Indian, Nest of, 85.
Tailors, 85.
Teal, American, points of distinction, 229.
Teal, American, The, 210.
Tell, The birds that remain in the, 244.
Temperature and nidification, 321.
Tern, Arctic, Nesting of, 116.
Tern, Black, points of distinction, 51
Tern, Black, The, 45.
Tern, Caspian, points of distinction, 212.
Tern, Caspian, The, 210.
Tern, Common, Nesting of, 116.
Tern, Lesser, Nesting of, 116.
Tern, Lesser, Winter home of, 245.
Terns, 115.
Terns and Gulls, Gregarious instincts of, 238.
Tern, Sandwich, Nesting of, 116.
Terns, Moving south of, 179.
Tern, Whiskered, points of distinction, 51.
Tern, Whiskered, The, 45.
Tern, White-winged Black, points of distinction, 51.
Tern, White-winged Black, The, 45.
Thaw, The, 267.
Thrushes in gardens, 113.
Thrushes in the snow, 274.
Thrushes, Moulting of, 120.
Thrushes, Second broods of, 110.
Thrushes, Time of incubation in, 111.
Thrush, Missel, Nest of, 72.
Thrush, Rock, points of distinction, 47.

INDEX. 351

Thrush, Rock, The, 40.
Thrush, Song, Autumn song of, 170.
Thrush, Song, Nest of, 19, 71.
Thrush, Song, Pairing of, 293.
Thrush, Song, Return of, 290.
Thrush, Song, Winter song of, 290.
Thrush, Song, Winter migration of, 277.
Thrush, White's, points of distinction, 215.
Thrush, White's, 198.
Thrush, White's, Colour of wing of, 250.
Thunder showers, Birds in, 107.
Tit, Continental Long-tailed, points of distinction, 217.
Titmice, Footprints of, 305.
Titmice, Habits of, in Winter, 275, 287.
Titmice, Nests of, 21, 78, 86.
Titmice, Pairing of, 293.
Titmice, Roosting-places of, 327, 331.
Titmouse, Bearded, Habits of, 182.
Titmouse, Great, Nest of, 80.
Titmouse, Lapp, Plumage of, 255.
Titmouse, Long-tailed, Arctic form of the, 200.
Titmouse, Long-tailed, Nest of, 80.
Tits, Habits of, in Summer, 122.
Totanus, Footprints of, 301.
Tringa, Footprints of, 300.
Turnip fields in Autumn, The, 175.
Turtle Dove, date of laying, 102.
Twite, date of flocking, 121.
Twite, Disbanding of, 13.
Twite, Note of, 30.
Twites, Roosting-place of, 330.

Universality of life, 99.
Upholsterers, 84.

Vulture, Egyptian, points of distinction, 213.
Vulture, Egyptian, The, 196.
Vulture, Griffon, The, 196.
Vulture, Griffon, points of distinction, 213.
Vultures, Power of scent in, 67.
Wagtail, Blue-headed, points of distinction, 48.

Wagtail, Blue-headed, The, 42.
Wagtail, Gray, Nest of, 19.
Wagtails, Roosting-place of, 333.
Wagtail, White, points of distinction, 48.
Wagtail, White, The, 42.
Wagtail, Yellow, Winter home of, 244.
Wagtail, Yellow, Young of, 106.
Warbler, Aquatic, points of distinction, 47.
Warbler, Aquatic, The, 41.
Warbler, Barred, points of distinction, 47, 48.
Warbler, Barred, The, 41, 46.
Warbler, Blue-throated, The, 199.
Warbler, Garden, Arrival of, 16.
Warbler, Garden, Winter home of, 244.
Warbler, Grasshopper, Habits of the, 104.
Warbler, Grasshopper, Winter home of, 242.
Warbler, Great Reed, points of distinction, 47.
Warbler, Great Reed, The, 41, 46.
Warbler, Icterine, points of distinction, 47.
Warbler, Icterine, The, 41.
Warbler, Marsh, Winter home of, 244.
Warbler, Orphean, points of distinction, 48.
Warbler, Orphean, The, 41.
Warbler, Reed, Departure of, 184.
Warbler, Reed, Habits of, 108, 109.
Warbler, Reed, Nest of, 83.
Warbler, Reed, The, 46.
Warbler, Reed, Winter home of, 243, 244.
Warbler, Rufous, points of distinction, 217.
Warbler, Rufous, The, 199.
Warblers, Arrival of, 16.
Warbler, Sedge, Winter home of, 244.
Wash, The, in Autumn, 180.
Wash, Migration over the, 181.
Waterhen, Nests of, 108.
Waxwing, Distribution of, 310.
Waxwing, Flight of, 311.

Waxwing, habits in the snow, 311.
Waxwing, Nest and eggs of, 312.
Waxwing, points of distinction, 218.
Waxwing, The, 201.
Waxwing, Visits of the, 278.
Waxwing, Young of, 313.
Weasel chasing rabbits, 307.
Weaver Birds, Nests of, 81.
Weavers, 80.
Wheat, Among the, 143.
Wheatear, Arrival of, 11.
Wheatear, Departure of, 172.
Wheatear, Desert, points of distinction, 216.
Wheatear, Desert, The, 199.
Wheatear, Winter home of, 243.
Where the migrants go, 240.
Whinchat, Departure of, 172, 177.
Whinchat, Nest and eggs of, 24.
Whinchats, Note and habits of, 110.
Whinchats, Notes of, 23.
Whinchat, Winter home of, 242.
Whitethroat, Arrival of, 16.
Whitethroat, Common, Winter home of, 244.
Whitethroat, Nest of, 82.
Wigeon, American, points of distinction, 229.
Wigeon, American, The, 210.
Willow Wren, Arrival of, 12.
Willow Wren, Song of, 57.
Willow Wren, Song of, after moult, 121.
Willow Wren, Winter home of, 242.
Willow Wren, Yellow-browed, points of distinction, 217.
Willow Wren, Yellow-browed, The, 200.
Wings, Display of colour on, 253, 254.
Winter, 263.
Winter, Among the birds in, 272.
Winter and Spring, The struggle between, 270.

Winter and Summer range, Overlapping of, 243.
Winter, Calendar for, 335.
Winter, Habits of birds in, 322.
Winter morning, Habits of birds on, 287.
Winter quarters of migratory birds, 240.
Winter quarters, Unknown, 242.
Winter, Some birds of the, 309.
Winter, The terrors of the, 263.
Wolley and the Waxwing, 312.
Woodcock, Footprints of the, 305.
Woodcock, Nesting of, 22.
Woodcocks, Arrival of, 175.
Woodcocks, Solitary habits of, 281.
Wood Cutters, 77.
Wood Larks, 285.
Woodpecker, Green, Nest of, 77.
Woodpeckers, Habits of, in Winter, 276.
Woodpeckers, Tracks left by, 305, 306.
Wood Pigeons, Autumn flights of, 174.
Wood Pigeons, Roosting of, 174.
Woods, The, in Autumn, 175.
Wood Wren, Arrival of, 17.
Wood Wren, Departure of, 177.
Wood Wren, Winter home of, 243.
Wren, Autumn song of, 170.
Wren, Favourite roosting-place of, 329, 331.
Wren, Late building of, 119.
Wren, Nest of, 80.
Wren, Song of, 18.
Wryneck, Note of, 21.
Wryneck, Shamming death by, 128.
Wryneck, Winter range of, 243.

Yellow-browed Willow Wren, 200, 217.
Yellow Bunting, Late broods of, 120.
Yellow Bunting, Pugnacity of, 57.
Yellow Bunting, Song of, 192.
Yellow Wagtail, 106, 244.

Zones of winter quarters, 240.

www.ingramcontent.com/pod-product-compliance
Lightning Source LLC
Chambersburg PA
CBHW032046220426
43664CB00008B/881